Earthquake Engineering

WILEY SERIES IN

METHODS AND APPLICATIONS IN CIVIL ENGINEERING

Edited by
James R. Pfafflin
Department of Civil and Environmental Engineering,
New Jersey Institute of Technology,
Newark,
New Jersey
USA

EARTHQUAKE ENGINEERING
S. F. Borg

Earthquake Engineering

Damage Assessment and Structural Design

S. F. Borg

Professor of Civil Engineering
Stevens Institute of Technology
Hoboken,
New Jersey, USA

A Wiley Heyden Publication

JOHN WILEY AND SONS
Chichester · New York · Brisbane · Toronto · Singapore

Copyright © 1983 by Wiley Heyden Ltd.

All rights reserved.

No part of this book may be reproduced by any means, nor transmitted, nor translated into a machine language without the written permission of the publisher.

Library of Congress Cataloging in Publication Data:

Borg, Sidney F.
 Earthquake engineering.

'A Wiley Heyden publication.'
 Includes index.
 1. Earthquake engineering.   I. Title.
TA654.6.B67   1983   624.1'762   83-1304

ISBN 0 471 26261 7

British Library Cataloguing in Publication Data:

Borg, S. F.
 Earthquake engineering.
 1. Earthquake engineering
 I. Title
 624.1'762   TA654.6

ISBN 0 471 26261 7

Filmset Monophoto Times by Mid-County Press, London SW15
Printed by Page Bros. (Norwich) Ltd

This book is dedicated to the women in my life.

*First and foremost,
Audrey, my dearest wife*

Then, my daughter, Jill

And my daughters-in-law, Lise and Susan

And my grand-daughters, Kristina and Laura and Margo

And my mother, Pauline

Contents

Chapter	Page
Preface	ix
Introduction	xi
1 A Deep Focus Earthquake Mechanism	1
2 The Accelerogram Invariant and Its Parameters	14
3 The Isoseismal Invariant and Its Parameters	28
4 Earthquake Engineering Design Charts	41
5 Approximate Analytical Damage Assessment Procedures	57
6 Special Topics in Earthquake Structural Engineering	63
(a) Equivalent length and equivalent base area	64
(b) Length-of-time effect	65
(c) Damping of vibrational effects due to an earthquake	67
(d) Model — scaling requirements	68
7 The Structural Analysis Procedures: Symmetry, Anti-symmetry, Energy	77
Appendix Earthquake engineering and applied mechanics	103
Index	107

Preface

A new approximate approach to various aspects of earthquake engineering is presented in this book. The theory developed will be applicable to earthquakes of, roughly, magnitudes greater than five.

The fundamental quantity — the key to all of the procedures and methods derived — is energy. Starting with the initiation of the earthquake (a mechanism) and proceeding through timewise and spacewise analyses of energy on the surface of the earth, one is led to a procedure, based upon simple energy principles, for an approximate analysis of structures within the effective earthquake field. Two major observables — two fundamental sets of field or experimental data — are utilized in the theoretical developments. These two sets of data are: 1. The accelerograph record which gives timewise information, at a point, about the earthquake; and 2. The isoseismal contour chart which gives spacewise information over the region affected by the earthquake.

Invariants are looked for and obtained. That is, approximate relations or equations or curves are developed that hold for all accelerograms and for all isoseismal charts, within an accuracy consistent with the manner in which these data are obtained and also consistent with the engineering applications for which these are used.

New fundamental parameters are introduced — terms that are strongly related to the earthquake event and only to the earthquake event. Indeed, it was the author's conviction that this must be so which led him to the various parameter–invariant relations. These, in turn, by a series of logical rational steps assume a form which is of engineering use. This is done by extending the form of the invariant quantities and transforming them into relations that determine temporal and spacewise variations of surface energy.

This surface energy is related to damage and finally is utilized as the basis for the structural analysis process.

Engineering simplicity is looked for but not at the expense of accuracy suitable for ordinary engineering purposes. A major aim was to develop an approximate comprehensive rational earthquake engineering theory that could be used by everyday engineering design offices using elementary computer programs. The book will succeed or fail as it is or is not so used.

Three major sources of support must be thanked and acknowledged:
1. The National Science Foundation for their seed grant PFR 7822846 which enabled the author to develop the initial ideas discussed in the text.
2. The Trustees of Stevens Institute of Technology for granting the author a

sabbatical leave during which time the book could be written.

and finally (but not the least),

3. Professor Dr. John H. Argyris, Director of the Institute of Statics and Dynamics of Aero-Space Structures at the University of Stuttgart for inviting the author to be a Guest Professor at his Institute during the sabbatical year and for there providing the facilities and atmosphere so necessary and helpful in the preparation of the manuscript.

Introduction

In 1976, the National Science Foundation-Department of Interior (USGS) issued a detailed plan with options for augmenting the earthquake related research programs of various government agencies.* The plan lists a number of critical areas of required research.

For example, in *Sec. 5 Engineering, Objectives and Activities*, it describes several desirable (and perhaps necessary) areas of research that require clarification before engineers can really hope to develop rational damage assessment and structural analysis–design procedures.

Among those mentioned which relate to the material in this text-book are the following:

Subelement a: Characterization of ground motion for structural analysis and design
Objective: Develop methods to characterize the nature of the input motions and corresponding response of simple systems for use in engineering analysis, planning and design.
Activities: 1. Develop analytic models to estimate the special characteristics of ground motion and the acceleration, velocity and displacement time-histories of this motion for use as input motion in structural analysis and design.
2. Develop techniques for measuring the severity of earthquake effects based on parameters significant in engineering analysis and design.
Subelement d: Investigation of structural response
Objective: Develop analytical procedures for characterizing the earthquake response of structures and structural elements based on both analytical and experimental studies
Subelement f: Post-earthquake investigations
Objective: Obtain information for engineering analysis and design from observations of damage (or lack of damage) following earth-

---

* *Earthquake Prediction and Hazard Mitigation Options for USGS and NSF Programs*, Sept. 15, 1976. A more recent study, *Earthquake Engineering Research—1982*, prepared by Committee in Earthquake Engineering Research, Commission on Engineering and Technical Systems, National Research Council, published by National Academy Press, Washington, DC 1982, generally repeats the recommendations of the earlier report.

quakes that support the development of improved U.S. engineering practices and construction techniques.

This text presents rational approaches that bear directly on the important topics described above.

There are three basic principles upon which the theories are founded:

1. Earthquake engineering is a unique discipline in applied mechanics and as such has its own particular invariants, parameters, variables, and similar quantities.
2. The source for all the quantities mentioned in 1. are the two major observation banks (or experimental data or field data) of earthquake engineering, these being:
   (a) The accelerogram which, physically, must be related to the variation with *time* of ground energy at a *point* in the earthquake field.
   (b) The isoseismal contour map which, physically, must be related to the variation with *distance* of the ground energy over the *entire area* affected by the earthquake.

Therefore:

3. *Energy* is the key element in the earthquake event, starting from its initiation (the mechanism) and proceeding timewise and spacewise until its completion.

Equations are obtained for the time–energy and space–energy variations, based upon reasonable physical–technical hypotheses and a study of canonical accelerograms and isoseismal maps. Geological and frequency effects are included in an approximate manner, as are the soil–foundation interactions. All are given in terms of the newly introduced parameters, the 'acceleration index' and the 'isoseismal index' and a first approximation of design charts is given. These relate — in a form suitable for engineering office use — all of the elements that must appear in an engineering design analysis, namely magnitude of the earthquake, efficiency of the earthquake, geology, acceleration index, isoseismal index, soil–foundation interaction, and location–geometry–construction of the structure.

The rational assessment of damage is directly related to the new parameters and invariants. The structural design procedure also utilizes these quantities as well as symmetry — anti-symmetry considerations combined with a free–free beam type vibrational analysis. As part of the overall study, an analysis of model-prototype requirements and how these relate to current testing procedures is developed and critically examined.

In a very general way, the material included in the text covers the entire earthquake event, starting from its initiation (the 'mechanism') up to and including the effect of this earthquake on a structure.

Furthermore, the entire event is treated by a theory which is consistent with the ground rules stated above. In other words — we begin with a theoretical mechanism, for deep focus earthquakes. Following this, the accelerogram and isoseismal invariants are derived, their properties analyzed and their connections with energy developed.

This theoretical basis is then utilized in connection with the two main problems considered in this book, as noted above and as indicated in the text sub-title.

Overall, the treatment presented represents an approximate, rational, comprehensive, theoretical–applied approach to the problems. The basic experimental data — the accelerogram and isoseismal chart — are woven into a single unified theoretical framework which permits the ordinary engineering design office to determine the damage likelihood and also the structural response (shear, moment, deflection) of a particular structure at a particular location when subjected to a particular earthquake.

And all of the above is given in terms of elementary mathematical, physical and engineering concepts and lends itself to computer formulation.

1

A Deep Focus Earthquake Mechanism

INTRODUCTION

A mathematical–physical model of a deep-focus earthquake will be derived. The model, which is a 'mechanism', is a mathematical formulation that may apply to earthquakes that originate within the mantle of the earth, say 50–800 km below the surface of the earth.

This derived mechanism will have the following properties:

1. It will be a complete, closed form solution to the field equations and boundary conditions assumed for the event.
2. It will be based upon a reasonable, realistic, possible physical explanation for the development of the earthquake.
3. The instability, or trigger, for the initiation of the earthquake is included as a fundamental element in the solution.
4. It will account for the enormous amount of energy which, it is estimated, is released in a major earthquake. This energy shall be calculable using approximate relations.
5. It will account for the production of P and S waves that are generated in earthquakes.
6. It will predict a number of phenomena that may be subject to checks by geologists, rock scientists, and others working in this field.

The proposed mechanism requires the generation of very high stresses by ground surface standards. Although there may be doubt as to whether these could be generated on the surface of the earth, it is a fact that within the mantle the rock is subjected to enormous pressures and these affect the properties of the material in some, as yet unknown, ways. Thus, the required stresses and other properties (including the initiation of the triggering fracture, possibly by some phase change) may conceivably occur as required for the solution presented. Only a check of actual phenomena in the field can prove or disprove this point.

It is certainly a fact that enormous amounts of energy are released in earthquakes. Since these are probably generated by the release of 'locked-in' strain energy, it is conceivable that very high stresses acting on large volumes of material are involved. Both the high stresses and large volumes are part of the mechanism being discussed.

It seems clear that different tectonic earthquakes are caused by more than one mechanism. There is little doubt that fault slippage, such as occurs in typical California quakes, is a common accompaniment of many quakes. However, as noted by Newmark and Rosenblueth,[1] some seismologists hold that earthquakes originate in phase changes of rocks, rather than by fault slippage. Those who favour the phase change (volume change) theory argue that there is little likelihood that geological faults exist below depths of a few hundred kilometres because of the high temperatures and confining pressures, and yet data have been interpreted to indicate that earthquakes have originated at depths exceeding 600 km and up to 800 km.

An article in *Science*[2] reports on a meeting of geologists, seismologists, and engineers held in Knoxville, Tennessee in September, 1981. Among other activities, various speculations concerning possible mechanisms were presented. Quoting from the reference: 'John Armbruster and Leonardo Seeber of Lamont–Doherty Geological Observatory, in particular, in discussing the major 19th-century earthquake in Charleston, South Carolina, favoured a nearly horizontal fault separating an upper thin sheet of rock from the crust beneath the fault. They argue that the most violent effects of the 1886 Charleston earthquake covered too large an area to have resulted from a break on a nearly vertical fault. A break on a horizontal fault, on the other hand, could have directed its seismic energy over a much larger area of surface. A horizontal break could be caused by a tendency of the thrust sheet to backslide off the continent...'

The mechanism developed in this chapter is generally consistent with the Armbruster–Seeber view, although it is not a unique representation or solution for the assumed phenomena. It will, however, represent a broad, particular explanation of an earthquake which, by extension and generalization, may give some insight and knowledge of the actual details behind the build-up to and occurrence of a deep-focus earthquake. In this way, some important leads and hints may emerge relating to the two most important problems in earthquake engineering, still unsolved:

1. How to predict when an earthquake will occur.
2. How to 'defuse' an earthquake that is about to occur.

NOMENCLATURE FOR THIS CHAPTER

C = velocity of small disturbance in the solid (velocity of sound)
p = pressure of superplastic material
P = pressure on ruptured area material
r = radial distance coordinate
t = time
u = particle velocity
U_I = initial strain energy of element
U_F = final strain energy remaining in element following rupture
ξ = similarity coordinate
θ = angular variation

ρ = density
σ = stress, positive when tension
v = Poisson's ratio
μ = viscosity
o = subscript, outer
i = subscript, inner

PHYSICAL ASSUMPTIONS

Deep within the mantle, where the earthquake is assumed to originate, we have a condition of hydrostatic stress, which is shown below on a membrane or strip of thickness t. We assume this is a zero, initial condition, just as such a strip on the surface of the earth is subjected to a hydrostatic atmospheric pressure. Due to tectonic plate movement of the crust of the earth mantle, a tensile stress σ_o is added to the above strip of mantle (see Figures 1.1 and 1.2). This stress is assumed

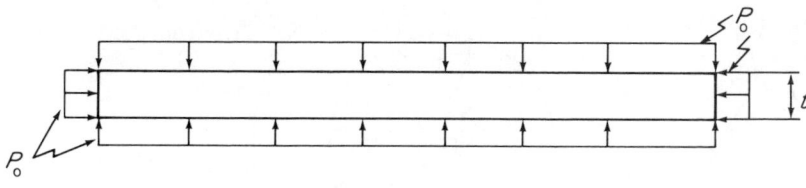

Figure 1.1

to be uniform along the edge of an enclosed area of indefinite extent and we consider the effect of σ_o alone. Thus the layer, initially subjected to the enormous hydrostatic pressures, now has the stresses σ_o applied to it uniformly around the boundary and these stresses introduce a strain energy in the layer caused by stretching (see Figure 1.2). The stress, σ_o, gradually increases and builds up as the tectonic plate or other movement action proceeds. Finally, a value of σ_o is reached that, in conjunction with the initial state of the layer and a possible phase change, leads to a 'rupture' or 'fracture' initiated at a single point. A (see Figure 1.3). (It is interesting to note that Benioff[3] has suggested, on the basis of observed wave forms from three earthquakes, that a class of deep earthquakes may arise when a

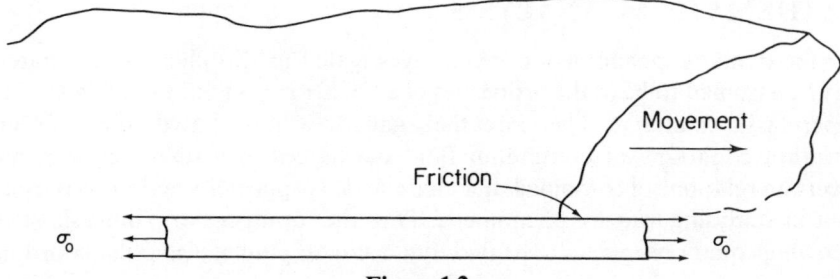

Figure 1.2

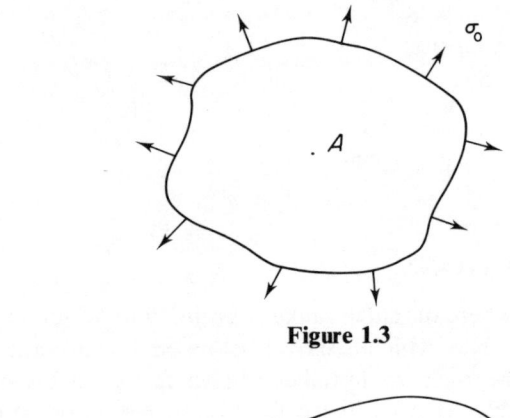

Figure 1.3

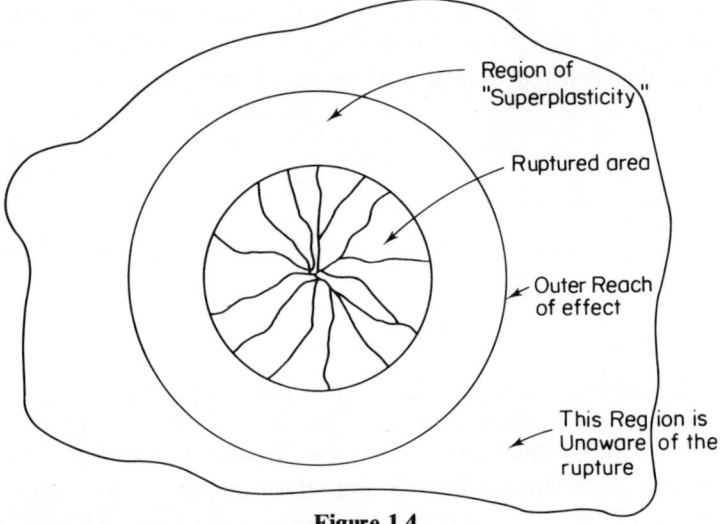

Figure 1.4

phase transition occurs through a region of rock, causing a sudden change in volume.) This failure at point A spreads radially throughout the layer, so that at a time t after initiation of the rupture conditions are as shown in Figure 1.4.

The mathematical development governing the event follows.[4]

MATHEMATICAL ANALYSIS

For the dynamic phenomenon being investigated in this chapter, the material may be assumed to have the properties of a border region fluid–solid, a so-called 'superplastic' material. Therefore the analysis will be based upon the conservation equations of continuum fluid mechanics instead of the somewhat uncertain relations of combined dynamic elasticity–plasticity action, as is usually done in analysing fracture phenomena. Thus the rupturing strip must satisfy the following equations (given in the two-dimensional symmetrical polar coordinate form).

Mass conservation:

$$\frac{\partial \rho}{\partial t} + \frac{\partial (\rho u)}{\partial r} + \frac{\rho u}{r} = 0 \qquad (1)$$

Momentum conservation (Navier–Stokes equation):

$$\frac{\partial u}{\partial t} + u\frac{\partial u}{\partial r} = -\frac{1}{\rho}\frac{\partial p}{\partial r} + 4/3\mu\left(\frac{\partial^2 u}{\partial r^2} + \frac{1}{r}\frac{\partial u}{\partial r} - \frac{u}{r^2}\right) \qquad (2)$$

Also, an equation for the velocity of small disturbances (since the effect of the point rupture spreads radially and reaches a finite distance from the centre of rupture),

$$C = C(p,\rho) \qquad (3)$$

In the foregoing equations, p is the negative mean value of the principal diagonal elements of the stress tensor, i.e.,

$$p = -1/3(\sigma_r + \sigma_\theta + \sigma_z) \qquad (4)$$

so that for the uniform strip stress field we have ($\sigma_r = \sigma_\theta$, $\sigma_z = 0$),

$$p = -\frac{2\sigma_r}{3} \qquad (5)$$

Physically, the phenomenon may be described in the following manner (this explanation will justify the assumed boundary conditions for the foregoing mathematical formulation of the field equations): (A) For $t < 0$, the entire strip is in a uniform tensile stress state, σ_o; (B) At $t = 0$, a small puncture is introduced at some interior station. We may think of this as the sudden introduction, along the arc of a small circle (the puncture point) of an equal and opposite, i.e. compressive, stress, $-\sigma_o$. Hence at this inner circle we have

$$\sigma_\theta = \sigma_{\theta F}$$
$$\sigma_r = 0$$
$$\sigma_z = 0 \qquad (6)$$
$$p = \frac{-\sigma_{\theta F}}{3}$$

$\sigma_{\theta F}$ is the 'fracture' or 'tearing' stress for the material, i.e. the tensile stress that will just cause the strip to fracture; and (C) For $t > 0$, conditions are as shown in Figure 1.5.

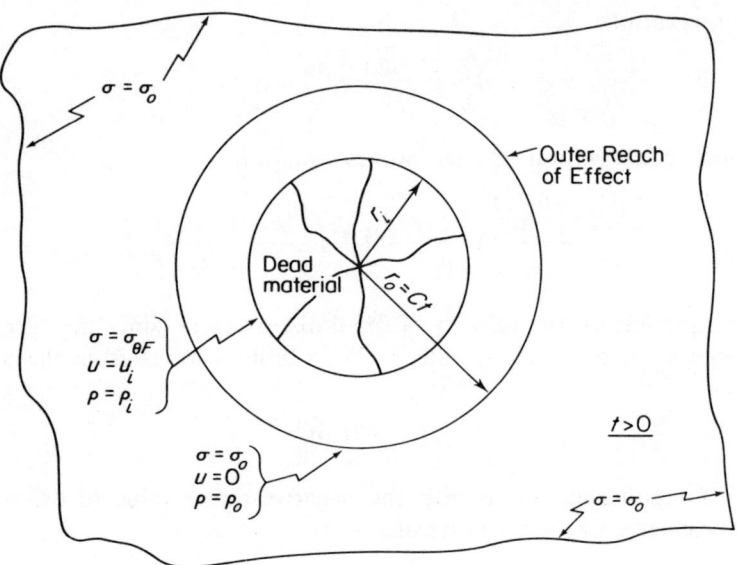

Figure 1.5

1. The punctured region now has a radius r_i moving with an assumed constant velocity, the velocity of fracture. The stress along the circle bounding this region is

$$\sigma_\theta = \sigma_{\theta F}$$
$$\sigma_r = 0$$
$$\sigma_z = -P_i \qquad (7)$$
$$p = -\frac{\sigma_{\theta F}}{3} + \frac{P_i}{3}$$

in which P_i is the final pressure on the entire ruptured strip (see Eq. 16 for the value of this stress) due to the swelling of the ruptured layer that has been relieved of stress. Furthermore, P_i almost certainly decays to zero in a boundary layer,[5] i.e. a very thin layer in the neighbourhood of r_i. Hence, it is not unreasonable to assume the region $r_i \leqslant r \leqslant r_o$ has $\sigma_z = 0$ and this will be done.
2. The material density at $r_i = \rho_i$
3. Within the radius r_i is a region of 'dead' material with cracks and fractures as required.
4. The outer reach of the effect is at a radius $r_o = Ct$. The particle velocity at r_o is zero and the stress at r_o is

$$\sigma_\theta = +\sigma_o$$

$$\sigma_r = +\sigma_o$$

$$\sigma_z = 0 \tag{8}$$

$$p = -\frac{2\sigma_o}{3}$$

The material density is ρ_o, the initial stressed strip density.
5. For $r_i \leqslant r \leqslant r_o$, the particle velocity, strip density, and stress condition have some variable values as required by the field equations and boundary conditions given above.

Because of the constant velocities of the outer and inner boundaries, and the constant stress $\sigma_{\theta F}$ at the inner boundary, an essential and fundamental simplification is involved by introducing the similarity coordinate

$$\xi = r/t \tag{9}$$

Since (based upon dimensional considerations) u must be linear in ξ, it follows that when

$$\frac{\mu}{t\xi_o^2} \ll 1 \tag{10}$$

we may neglect the viscosity term in the Navier–Stokes equation. We will assume this is so, and we shall neglect this quantity in our subsequent analysis. Then Eqs. (1), (2), and (3) become:

Mass conservation:

$$-\xi\frac{d\rho}{d\xi} + \frac{d(\rho u)}{d\xi} + \frac{\rho u}{\xi} = 0 \tag{1a}$$

Momentum conservation:

$$-\rho\xi\frac{du}{d\xi} + \rho u\frac{du}{d\xi} = -\frac{dp}{d\xi} \tag{2a}$$

State:

$$C = \left(\frac{dp}{d\rho}\right)^{\frac{1}{2}} \tag{3}$$

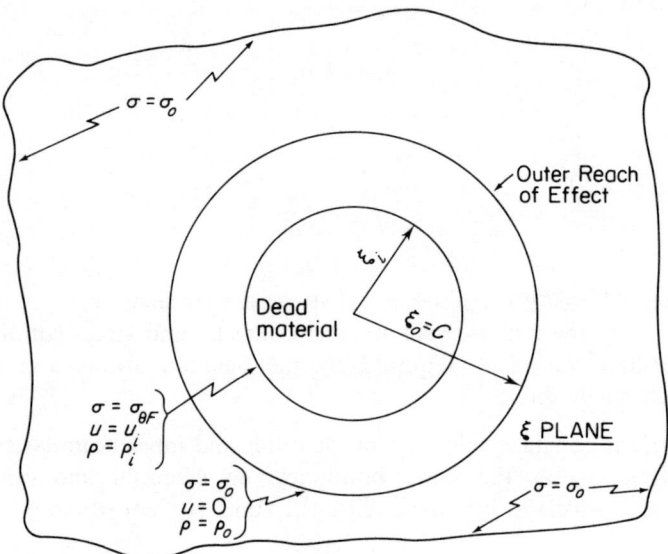

Figure 1.6

Now, the phenomenon which is represented by a different map in the physical (r, t) plane for each time t, is shown by a single field in the ξ-plane which holds for all times t, Figure 1.6. Note $\xi_o = C$ is the velocity of sound of small disturbances in the stressed strip and is the outer reach of affect; ξ_i is the velocity of the fractured boundary and hence also the fracture velocity in the strip.

The solutions to these equations, which satisfy all of the physical and boundary conditions as previously given in (A), (B), and (C), are

$$u = \xi_o - \xi \tag{11}$$

$$\rho = \rho_o \frac{\xi_o}{\xi} \tag{12}$$

$$2\sigma_o + 3p = 6\rho_o \xi_o (\xi_o - \xi) - 3\rho_o \xi_o^2 \ln \frac{\xi_o}{\xi} \tag{13}$$

On physical grounds, Eq. (11) and (12) imply that the velocity of the inner fracture surface is very likely nearly equal to the velocity of sound in the stress strip. In this connection, it is interesting to note that Mott[6] found that the velocity of travel for a linear fatigue crack in metals is given by approximately

$$\xi_i = 0.38 \xi_o \tag{14}$$

as compared to the value

$$\xi_i \sim C \tag{15}$$

assumed in the foregoing.

Also, a value for velocity of rupture of about 2 to 3.5 km s^{-1} has been inferred for a few earthquakes from seismograms,[7] although these are probably slip-fault ruptures.

The solutions of Eqs. (11), (12), and (13) imply certain restrictions on the field quantities for the particular solution obtained:

1. As noted above, ξ_i is very nearly equal to ξ_o.

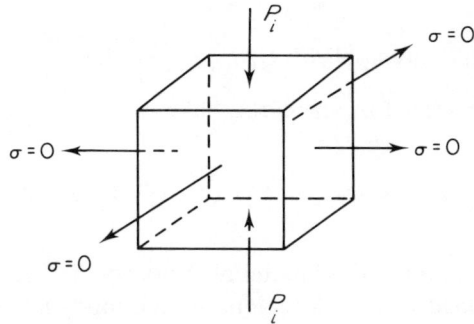

Figure 1.7

2. The conditions of the inner, ruptured material may be represented as in Figure 1.7, where a unit cube is shown. P_i causes a unit deflection just equal to that caused by σ_o in Figures 1.2 and 1.3. Therefore

$$P_i = 2v\sigma_o \tag{16}$$

and from Eq. 13 and the boundary conditions

$$\sigma_o = \frac{3\rho_o \xi_o^2}{(1+v)} \left[1 - \frac{\xi_i}{\xi_o} - \frac{1}{2} \ln \frac{\xi_o}{\xi_i} \right] + \frac{\sigma_{\theta F}}{2(1+v)} \tag{17}$$

This is the stress compatibility condition for the given solution.

The gross energy of the earthquake is just the net strain energy released in the 'dead material' layer. We may obtain an approximate expression for this in terms of σ_o, P_i and the properties of the dead material layer as follows, using the equations of elasticity

U_{INITIAL} = strain energy in the unruptured layer

$$= \frac{\sigma_o^2}{E}(1-v)(\text{volume}) \tag{18}$$

After failure, the dead material has

$$\sigma_z = P_i = 2v\sigma_o \text{ (compression)}$$
$$\sigma_o = 0 \tag{19}$$

and the strain energy

$$U_{\text{FINAL}} = \frac{2v^2\sigma_o^2}{E} \text{ (volume)} \tag{20}$$

(If P_i is assumed 'suddenly applied', $U_{\text{FINAL}} = \frac{4v^2\sigma_o^2}{E}$.)

Then the energy released by the earthquake is

$$U_{\text{EARTHQUAKE}} = U_{\text{INITIAL}} - U_{\text{FINAL}} = \frac{\sigma_o^2}{E}(1 - v - 2v^2)(V_{DM}) \tag{21}$$

In which V_{DM} is the volume of dead material. A portion of this energy is then 'lost' in fracturing the dead material region, in thermodynamic and, probably, chemical reactions as well.

The earthquake mechanism is then as shown in Figure 1.8. The fracture or rupture, caused when σ_o reaches the value given in Eq. 17, is a sudden one brought about by the extremely high stress and (possibly) a sudden phase change at a point. The dead material releases net strain energy and develops a pressure $P_i(=2v\sigma_o)$ which, very likely, pulsates as it pushes against the overlaying and underlaying material, introducing P and S waves and transfers the energy released by the fractured layer to, ultimately, the surface region affected by the earthquake.

To indicate approximate values for energy and deflection of P_i, if we assume the

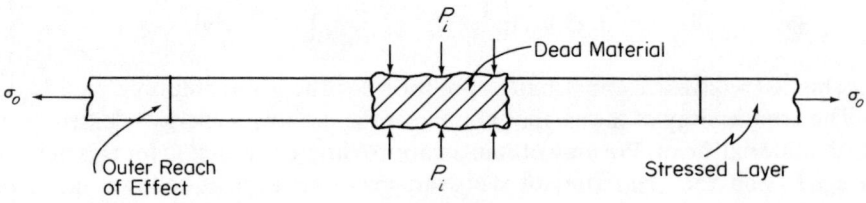

Figure 1.8

fractured area is 5000 ft in diameter and 100 ft thick, and using assumed values that follow:

$$\sigma_o = 200{,}000 \text{ psi}$$
$$P_i = 2v\sigma_o = 120{,}000 \text{ psi}$$
$$E = 3 \times 10^6 \text{ psi} \tag{22}$$
$$v = 0.3$$

Assume the energy lost in fracturing $= 10$ per cent $(U_I - U_F)$, we find

$$U_{\text{EARTHQUAKE}} = 2 \times 10^{15} \text{ in lb} \tag{23}$$

and a maximum deflection of P_i of about 10 in.

The energy value obtained is not unreasonable.[1] Also, following the initial rupture, the enormous pressures 'heal' the fracture region. More movement and phase change in a vulnerable or weakened layer may be an explanation for aftershocks.

TO BE DONE

Insofar as the proposed model is concerned, a number of future activities are indicated:

1. Values of the physical constants that occur should be determined. These will have to come from some of the centres working in deep-earth research, such as the one, for example, at Cornell University.[8]
2. Is there any evidence — geologic, seismograph, accelerogram, observations — that tends to confirm or contradict the proposed model, for deep-focus earthquakes?
3. Among the behaviours to look for are:
    (a) A 'pulsating' loading due to P_i, with the corresponding P and S waves (Figure 1.8).
    (b) An enormous release of energy in a matter of seconds (Eq. 22–23 and the assumed 5000 ft diameter ruptured area).
    (c) Energy release from either a layer or area a mile (miles) in extent or a thicker strip of lesser area.
    (d) Surface movement which leads to the build-up of the stress σ_o.
    (e) A phase change or other unusual behaviour or configuration of the material which, in conjunction with σ_o, results in a triggering instability-fracture or rupture.
    (f) A 'healing' of (e) due to the high pressures involved.
    (g) Any clues in the proposed mechanism that would lead to a prediction of a possible earthquake.

(h) Any means for triggering or defusing an earthquake due to the suggested mechanism.
4. Finally, can the proposed mechanism be connected with the Richter scale?

CONCLUSION

A mathematical-physical model of a particular deep-focus earthquake mechanism was derived. The model utilized equations of fluid flow as well as those of linear elasticity theory and a basic similarity transformation. In a sense, the analysis (which leads to a closed form solution) assumes a so-called 'superplastic' material.

Summarizing the key physical requirements, hypotheses, and results for the proposed mechanism:

1. A plate subjected to a uniform tensile stress field.
2. A critical value for this stress, σ_o.
3. A puncture or a phase change or an occlusion or whatever, at any rate, a rupture at a point A.
4. Which causes
    (a) A circular region centred at A of zero in-plane stresses surrounded by
    (b) A circular ring subject to the conservation equations of mechanics, in terms of r,t. These equations can be transformed by introducing the similarity coordinate $\xi = r/t$, so that the infinite number of time-dependent maps of the phenomenon in the r,t plane collapse to a single map on the ξ plane.
5. The rupture at point A causes a 'tearing or fracture' stress, $\sigma_{\theta F}$ at the inner boundary of 4(b), the rupture boundary, so that the rupture is self-sustaining and
6. The region 4(a) grows uniformly with time and this is the earthquake energy-producing region whose outer boundary moves with the constant velocity ξ_i, the rupture velocity.
7. ξ_i is also the constant velocity of the *inner* boundary of region 4(b). The outer boundary of this region moves with the constant velocity ξ_o, which is the velocity of sound in the stressed region, i.e., the velocity of small disturbances.
8. On physical grounds, because $u_i = \xi_o - \xi_i$

$$\frac{\xi_o}{2} < \xi_i < \xi_o$$

   so that
9. σ_o, the critical stress, which must be less than $\sigma_{\theta F}$ is given by

$$\frac{\sigma_{\theta F}}{2(1+v)} < \sigma_o < \sigma_{\theta F}$$

and therefore

$$\sigma_{\theta F} \geq \frac{3\rho_o \xi_o^2 (1 - ln2)}{1 + 2v}$$

The mechanism summarized above includes predictions of earthquake behaviour and parameters, *for the mechanism considered*, that may be checked, either directly or indirectly.

REFERENCES

1. Nathan M. Newmark and Emilio Rosenblueth. *Fundamentals of Earthquake Engineering*, Prentice Hall, Inc., Englewood Cliffs, N.J. 1071, Ch. 7.
2. Richard A. Kerr. Assessing the risk of Eastern U.S. earthquakes, *Science*, Vol. **214**, 9 Oct. 1981, p 169–171.
3. H. Benioff. Source wave forms of three earthquakes, *Bull. Seism. Soc. Am.*, **53**, 893, 1963.
4. S.F. Borg. Rupture instability of plane membranes and solids, *Journal of Applied Mechanics*, (ASME), Sept. 1960.
5. S.F. Borg. A note on Boundary Layer Type Solutions in Applied Mechanics, Journal of Aeronautical Sciences, April 1950.
6. N.F. Mott. Brittle fracture of mild steel plates, *Engineering* (British), **164**, 1947; **165**, 1948.
7. Bruce A. Bolt. Causes of Earthquakes, in *Earthquake Engineering*, Robert L. Weigel, Coordinating Editor, Prentice-Hall, Inc., Englewood Cliffs, N.J. 1970, Ch. 2.
8. See Materials at ultrahigh pressures, *Engineering, Cornell Quarterly*, **14**, No. 1, Summer 1979.

2

The Accelerogram Invariant and its Parameters

INTRODUCTION

One of the more important and useful records of an earthquake is the accelerogram. This data, which is a printout of acceleration in a given direction, as a function of time, at a given locality, is utilized by earthquake scientists to assist in the analysis of the earthquakes that developed the particular accelerograms, and also in the design of structures for future earthquakes. Because of the important role played by the accelerogram in many areas of earthquake engineering, it was considered crucial that a unifying concept or invariant relation for this fundamental data be obtained. The invariant, to be most useful, should be one that is generated by a rigorous mathematical and physical hypothesis capable of being checked experimentally and, furthermore, should be one that can be utilized in extended analysis of earthquake engineering phenomena in ordinary engineering design offices.

It will be shown in this chapter that such an invariant for the accelerogram does indeed exist. Furthermore, this invariant will be used as the basis for several applications in this and in Chapters 5 and 7 as follows:

1. In this chapter, the invariant of the accelerogram will be derived and various properties of it will be discussed.
2. It will then be utilized in Chapter 5 in developing a postulated 'damage criterion'.
3. In Chapter 7, the invariant will be utilized as one of the basic tools in the proposed method of structural analysis in earthquake engineering.[1]

It must be emphasized that, as is true for practically every event that occurs in earthquake engineering, the hypothesized accelerogram invariant does not represent an exact relation. It can only be considered as an 'average' for typical earthquakes with a spread depending upon a number of parameters. One of these important parameters must almost certainly be the 'geology' of the region affected by the earthquake. Among other factors that we shall include in the general term 'geology' or 'geologic region' are (a) the depth of the focus, (b) the overall configuration (i.e. boundary of continent or mountainous or plains region), and (c) the frequencies of the accelerations.

Another basic assumption in the present accelerogram analysis is that we are dealing with 'canonical' accelerograms. By canonical we mean an accelerogram which conforms to the derived invariant relation. Many accelerograms do, in fact, have this property. Furthermore, analysis indicates that those accelerograms which do not directly conform to the canonical requirements can generally be given by a simple linear superposition of canonical forms and hence can still be utilized in the theoretical analyses based upon the canonical expression. More on this point later in the chapter.

The invariant analysis developed herein is reasonable and consistent with one of the fundamental characteristics of the earthquake phenomenon, namely that all phases of it, starting from its initiation (the mechanism) to its final destructive effect, are variable and appear to be of uncertain and erratic form. One can make very few absolutely positive statements in dealing with earthquake phenomena. One can only speak of 'averages', 'probabilities', and similar terms in attempting to obtain a basic understanding of a particular event, as is the case in this and later chapters.

One additional consideration is that the accelerograms and the relations derived from them in this chapter refer only to the *horizontal ground* accelerations and do not apply to accelerograms obtained in or on buildings or other structures, or to vertical accelerations.

NOMENCLATURE FOR THIS CHAPTER

Following are the terms used in this chapter. They will be defined also when they are introduced.

a = ordinate to accelerogram envelope
f = subscript final
k, K = dimensionless terms
n = dimensionless term
t = time
A, B = numerical constants
ε_{t_f} = SHE = the total surface horizontal energy per unit area supplied by the earthquake at the site where acceleration is recorded
T = SHE per unit area per unit time supplied by the earthquake at the site where the acceleration is recorded
θ = angle between a ray from focus to location of accelerograph and the direction of the accelerograph

MATHEMATICAL–PHYSICAL ARGUMENT

A typical accelerogram is shown in Figure 2.1. Note in particular the following properties of the accelerogram which are characteristic of all 'canonical' accelerograms.

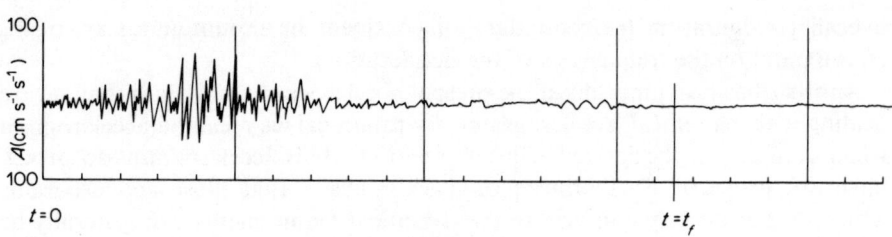

Figure 2.1

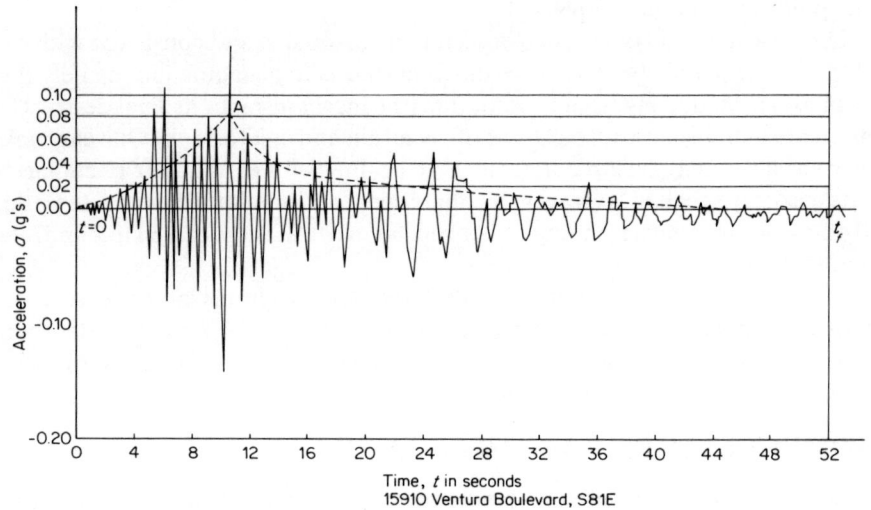

Figure 2.2

1. The acceleration starts at, essentially, zero corresponding to $t=0$. It then increases to a maximum and finally decreases to zero, corresponding to $t=t_f$.
2. The chart is approximately symmetric with respect to the time axis, so that positive and negative accelerations are assumed to be mirror images of each other.

In developing the accelerogram invariant, we form an envelope of the positive portion of the curve, as shown in Figure 2.2. The positive portion only is used because of (2) above, and in view of the following mathematical development and of the uses to which the accelerogram-envelope is put. In essence an approximate combined curve (positive and negative accelerations) is accounted for in what follows.

Starting at the initiation of acceleration increase ($t=0$) and proceeding to the final time (t_f) of acceleration change, and using the area under the envelope curve, we assume, considering two points, subscripts 1 and 2, a distance Δt apart,

$$\sum_{t=0} (a\,\Delta t)_2 = \sum_{t=0} (a\,\Delta t)_1 + \Delta(a\,\Delta t) \tag{1}$$

in which a is the ordinate (i.e. acceleration) to the envelope at the point considered. Eq. (1) is mathematically and physically reasonable.

The next statement is a key assumption in the mathematical formulation.[2] Based upon dimensional as well as physically reasonable arguments we assume that

$$\Delta(a\,\Delta t) = k \sum_{t=0} (a\,\Delta t)_1 \frac{t_f^n}{t^{n+1}} \Delta t \qquad (2)$$

in which k and n are non-dimensional constants whose values are to be determined.

Proceeding,[2] we obtain after integrating

$$\frac{\Sigma(a\,\Delta t)}{\Sigma(a\,\Delta t)_f} = e^{K1 - [t_f/t]^n} \qquad (3)$$

an equation which satisfies the initial and final conditions and which must be tested against the realities of actual accelerograms.

For clarity, we list again the definitions of the terms in Eq. (3).

$\Sigma(a\,\Delta t)$ = area under the accelerogram envelope between $t=0$ and $t=$ any time, t

$\Sigma(a\,\Delta t)_f$ = total area under the accelerogram envelope between $t=0$ and $t=t_f$

K, n = numerical constants to be determined

Eq. (3) is the postulated 'invariant' of the canonical earthquake accelerogram.

To test this hypothesis, the accelerograms of the following earthquakes were utilized:

(a) Tolmezzo, 1976[3]
(b) Taft, 1952[4]
(c) Lima, 1966[5]
(d) San Fernando, 1971[6]
(e) Bucharest, 1977[7]

It was found that a best fit for the data from the five earthquakes is given by (see Figure 2.3 and the Appendix to this chapter) the approximate relation

$$\frac{\Sigma(a\,\Delta t)}{\Sigma(a\,\Delta t)_f} = e^{0.12[1-(t_f/t)^{1.8}]} \qquad (4)$$

In Eq. 4, t_f and $\Sigma(a\,\Delta t)_f$ vary for each earthquake, and the values for these two quantities are listed in Table 2.1.

As pointed out above, K and n are constants. The significance of this is related to a postulated uniqueness–existence hypothesis which was formulated elsewhere[8] and the details of which will not be discussed further in this textbook,

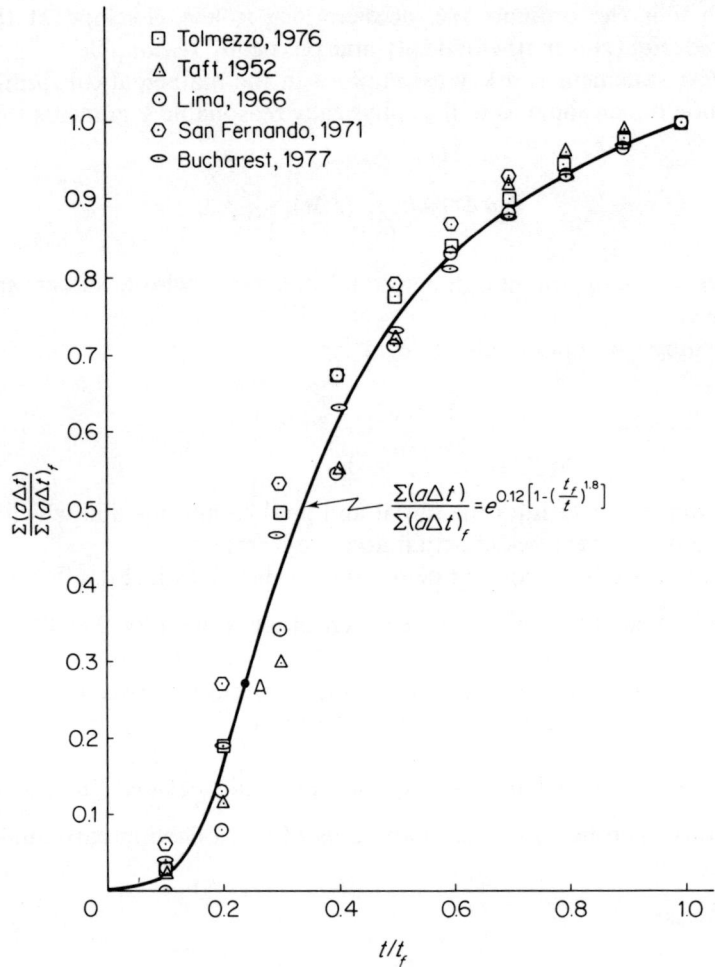

Figure 2.3

Table 2.1 t_f and $\Sigma(a\Delta t)_f$ for five earthquakes

Earthquake	t_f sec	$\Sigma(a\Delta t)_f$ 'g' sec
Tolmezzo, 1976	20	0.98
Taft, 1952	20	0.96
Lima, 1966	20	1.84
San Fernando, 1971	48	1.06
Bucharest, 1977	14	0.72

although the tentative conclusions obtained from it will be used later in this and the next chapter.

DETAILED ANALYSIS OF THE INVARIANT — EQ. 4

As indicated in Figure 2.3, a plot of the invariant, Eq. 4, is a typical S or growth curve. The point A, which is the point of zero curvature, corresponds to the point A on Figure 2.2 which is the maximum positive (and assumed negative) acceleration on the envelope curve of the accelerogram. This point occurs at approximately

$$\left(\frac{t}{t_f}\right)_A = 0.25 \tag{5}$$

and therefore, from Eq. 4,

$$\left.\frac{\Sigma\,(a\,\Delta t)}{\Sigma\,(a\,\Delta t)_f}\right|_A = 0.29 \tag{6}$$

The values given in Eqs. 4, 5, and 6 will at this time be assumed as fixed average values for typical canonical earthquake accelerograms subject to verification, possible change, or modification as a more detailed study and analysis of accelerograms throughout the world requires.

Once more it is emphasized that particular accelerograms, subject to special conditions, will not conform to the invariant form of Eq. 4. Thus, the accelerogram of the Pacoima Dam (Figure 2.4), which represents conditions almost directly over the focus, probably includes a number of shock-clusters and reflection–refraction effects and perhaps even dam vibration effects that do not, in general, occur in accelerograms some distance from the focus. However, the accelerogram of Figure 2.4 can very easily be approximated by the sum of two canonical acclerograms, as shown in Figure 2.5, indicating the likelihood of two

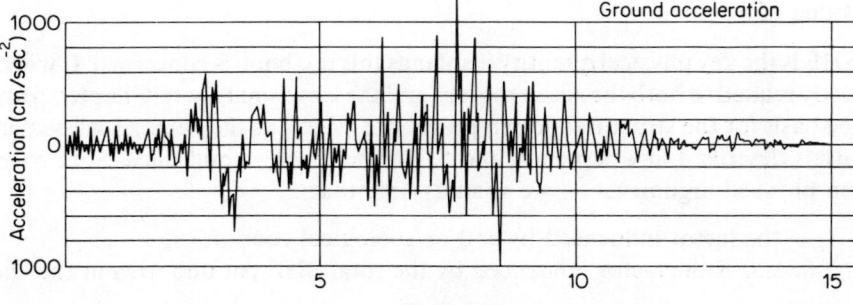

Figure 2.4

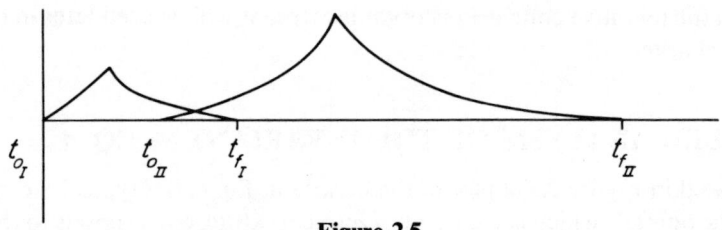

Figure 2.5

successive separate shocks occuring at and during the times shown. The combined effect may then be taken by considering the separate canonical accelerogram effects.

Finally, based upon the analysis here presented, it seems clear that the two fundamental parameters of the accelerogram are t_f and $\Sigma(a\,\Delta t)_f$, i.e. the total time of the accelerogram and the area under the envelope of the accelerogram.

THE PHYSICAL VARIABLES RELATED TO THE CANONICAL ACCELEROGRAMS

It was stated earlier in this chapter that a tentative uniqueness–existence hypothesis which related to Eq. 4 was developed.[8] According to this tentative hypothesis, there are two physical variables upon which the two fundamental parameters t_f and $\Sigma(a\,\Delta t)_f$ depend. An analysis of the earthquake phenomenon suggests that these two variables are:

1. Soil or geological conditions at the point where the accelerogram is obtained.
2. The total surface horizontal energy — SHE — per unit ground area developed by the earthquake at the site where the accelerogram is obtained. SHE is the energy which reaches the surface of the earth and which causes the horizontal vibration and damage of structures. In one form or another it is this quantity which must be determined if a rational method of structural analysis is to be obtained for structures subject to earthquake effects. SHE can be expressed in terms of an 'efficiency, η' in which η converts the total focal earthquake energy into the destructive surface energy. Bolt[9] discusses an efficiency to convert source energy to seismic energy but as he notes very little is known about the value of η.

SHE is the key physical quantity insofar as this textbook is concerned. It will be directly related to both the accelerogram and the isoseismal chart (Chapter 3) and is the basis for the structural design procedure as well as the damage assessment analysis described in Chapters 4 and 5. Furthermore — and once more based upon physical arguments — we shall assume that:

1a. t_f is the factor influenced by soil or geological conditions,
2a. $\Sigma(a\,\Delta t)_f$ is the factor influenced by the total SHE per unit area at the site.

Note that the form of the parameters indicates a possible cross-correlation.

Among the indicated topics for future research which this chapter suggests are studies to check the two major correlations hypothesized above.

As an approximation, we hypothesize that the determination of 'maximum accelerogram effect', i.e. $\{\Sigma(a\Delta t)_f\}_{\max}$, may be obtained from

$$\{\Sigma(a\Delta t)_f\}_{\max} = \frac{\{\Sigma(a\Delta t)_f\}_{\text{actual}}}{\cos\theta} \tag{7}$$

in which

$\{\Sigma(a\Delta t)_f\}_{\text{actual}}$ is the value obtained from a single available accelerogram, and

θ is the angle between a ray from the focus to the location of the accelerograph and the actual direction of the accelerograph.

Admittedly, Eq. 7 may be a poor approximation (or it may be a very good one) depending upon local soil conditions, and also upon geological features between the focus and the accelerograph. But these difficulties are inherent in the earthquake phenomenon. One looks for correlations and generalizations that apply 'on the whole'.

THE TIMEWISE VARIATION OF SHE AT A POINT

Eq. 4, which is an invariant of the accelerogram, enables one to obtain an approximate expression for the SHE as a function of time, supplied by the earthquake, at the ground site of the accelerogram. This is accomplished by postulating a parallel (although different) physically-based mathematical relation for the $\Sigma(a\Delta t)$ term which is assumed to be a basic parameter of the accelerogram record. The procedure is as follows. Assume

$$\frac{\Delta[\Sigma(a\Delta t)]}{\Delta t} = A\frac{T}{\varepsilon_{t_f}}\Sigma(a\Delta t) \tag{8}$$

in which A is a numerical constant*

ε_{t_f} is a physical constant related to the accelerogram response, say the total surface horizontal energy, SHE per unit area supplied by the earthquake at

---

* Concerning the parallel acceleration index representations: in Eq. 2, it is postulated that

$$\frac{\Delta\Sigma(a\Delta t)}{\Delta t} = f(\Sigma(a\Delta t), t)$$

The alternate formulation, Eq. 8 is

$$\frac{\Delta\Sigma(a\Delta t)}{\Delta t} = g(\Sigma a\Delta t, \text{SHE}(t))$$

In both the particular variables and parameters of earthquake engineering are used.
See Chap. 3, pp. 32 and 35 in which a very similar analysis is used for the isoseismal index. Note also the symmetries in the accelerogram–time and isoseismal–space in these and other equations and curves given in the text.

the site where the acceleration is recorded, and, based upon dimensional as well as physical requirements,

T is the derivative of SHE per unit area with respect to time, t, supplied by the earthquake at the site where the acceleration is recorded.

Now, comparing Eq. 8 above and Eq. 2, and equating equivalent terms, we find

$$T = B\varepsilon_{t_f} \frac{t_f^{1.8}}{t^{2.8}} \qquad (9)$$

in which B is a numerical constant. This expression, which has a singularity at $t=0$, will be analysed further in Chapter 4, and will be used in that and later chapters in the damage assessment and structural analysis procedures.

CONCLUSION

One of the major sources of field (experimental) data connected with earthquakes — the accelerograph record — is analysed with particular emphasis on the determination of the invariant relating thereto. Based upon a physical–mathematical–dimensional line of reasoning, a canonical accelerogram is postulated and an invariant for it is obtained. Two basic parameters are also postulated and define the invariant. These in turn are related to two physical variables of the earthquake event. Various properties of the canonical invariant are derived and the invariant itself is also related to the single most important quantity involved in the earthquake phenomenon — the energy. The equations, relations, and results obtained in this chapter will be utilized in later chapters dealing with damage assessment and also with structural design analysis.

REFERENCES

1. S.F. Borg. *Some New Approximate Procedures Relating to the Analysis of Structures Subjected to Earthquake Loadings — Part 1, Elastic Theory*. Technical Report ME/CE-80-1, Department of Mechanical Engineering/Civil Engineering, Stevens Institute of Tech., August, 1980.
2. S.F. Borg. *Similarity Solutions in the Engineering, Physical–Chemical, Biological–Medical and Social Sciences*, Proceedings of Symposium 'Symmetry, Similarity, and Group Theoretic Methods in Mechanics', P.G. Glockner and M.C. Singh (eds.), University of Calgary, August 1974.
3. CNEN–ENEL. *Commission on Seismic Problems Associated with the Installation of Nuclear Plants*, Contribution to the Study of Fruili Earthquake of May, 1976, Figure 4, p. 52.
4. Robert L. Wiegel. *Earthquake Engineering*, Figure 4.2, p. 78, Prentice-Hall, Inc., Englewood Cliffs, N.J. 1970.
5. Robert L. Wiegel. *Earthquake Engineering*, Figure 4.19, p. 89, Prentice-Hall, Inc., Englewood Cliffs, N.J. 1970.

6. I.M. Idress. *Characteristics of Earthquake Ground Motions*, Specialty Conference on Earthquake Engineering and Soil Dynamics, ASCE, Pasadena, California, June 19–21, 1978, p. 18.
7. Fattal *et al.* Observations on the Behaviour of Buildings in the Romania Earthquake of March 4, 1977, NBS SP 490, Sept. 1977.
8. S.F. Borg. *Generalized Tensors and Matrices*, Proc. Second International Conference on Mathematical Modeling, St. Louis, Missouri, July 1979.
9. Robert L. Wiegel (Ed.). *Earthquake Engineering*, Chap. 2 written by Bruce A. Bolt, Prentice-Hall, Inc., Englewood Cliffs, N.J. 1970.

APPENDIX

Tables 2.2 to 2.6 and the accompanying accelerograms show the values that were used in preparing Figure 2.3.

Table 2.2 Tolmezzo, 1976 (see Figure 2.6): $t_f = 20$ s, $\Sigma(a\Delta t)_f = 0.98$, $\cos\theta = 45°$

t/t_f	$\Sigma(a\Delta t)/\Sigma(a\Delta t)_f$	$e^{0.12[1-(t_f/t)^{1.8}]}$
0	0	0
0.1	0.03	~0
0.2	0.19	0.13
0.3	0.49	0.40
0.4	0.67	0.60
0.5	0.77	0.74
0.6	0.84	0.83
0.7	0.90	0.90
0.8	0.94	0.94
0.9	0.98	0.98
1.0	1.00	1.00

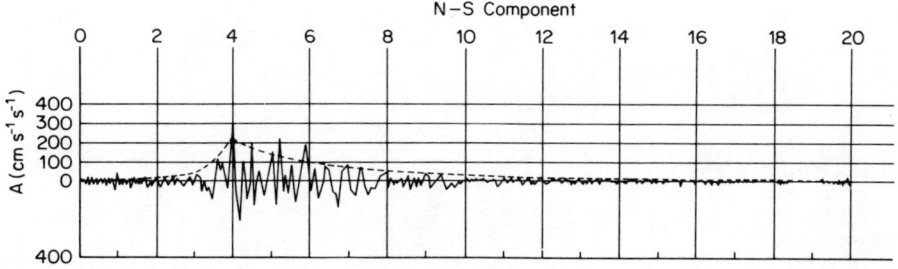

Figure 2.6

Table 2.3 Taft, 1952 (see Figure 2.7) $t_f = 20$ s, $\Sigma(a\,\Delta t)_f = 0.96$, $\cos\theta =$ Unknown

t/t_f	$\Sigma(a\,\Delta t)/\Sigma(a\,\Delta t)_f$	$e^{0.12[1-(t_f/t)^{1.8}]}$
0	0	0
0.1	0.03	~0
0.2	0.11	0.13
0.3	0.30	0.40
0.4	0.55	0.60
0.5	0.72	0.74
0.6	0.83	0.83
0.7	0.92	0.90
0.8	0.96	0.94
0.9	0.99	0.98
1.0	1.00	1.00

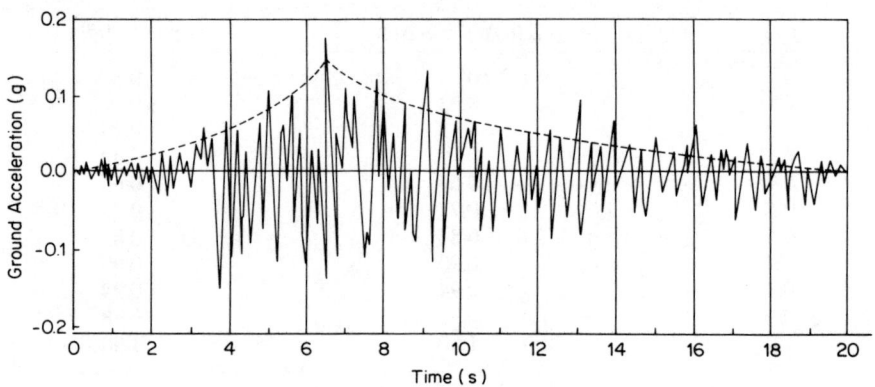

Figure 2.7

Table 2.4 Lima, 1966 (see Figure 2.8) $t_f = 20$ s, $\Sigma(a\Delta t)_f = 1.84$, $\cos\theta =$ Unknown

t/t_f	$\Sigma(a\Delta t)/\Sigma(a\Delta t)_f$	$e^{0.12[1-(t_f/t)^{1.8}]}$
0	0	0
0.1	0.03	~ 0
0.2	0.13	0.13
0.3	0.34	0.40
0.4	0.54	0.60
0.5	0.71	0.74
0.6	0.82	0.83
0.7	0.90	0.90
0.8	0.95	0.94
0.9	0.98	0.98
1.0	1.00	1.00

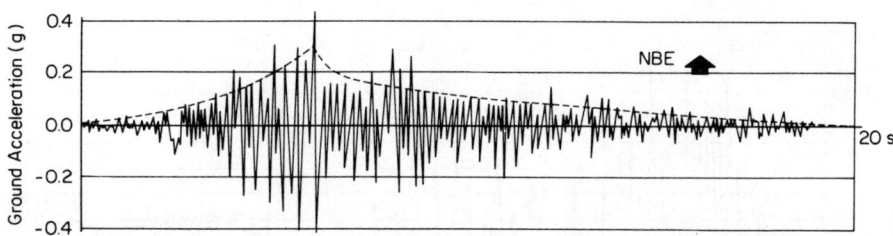

Figure 2.8

Table 2.5 San Fernando, 1971 (see Figure 2.9): $t_f = 48$ s, $\Sigma(a\,\Delta t) = 1.06$, $\cos\theta =$ Unknown

t/t_f	$\Sigma(a\,\Delta t)/\Sigma(a\,\Delta t)_f$	$e^{0.12[1-(t_f/t)1.8]}$
0	0	0
0.1	0.06	~0
0.2	0.27	0.13
0.3	0.53	0.40
0.4	0.67	0.60
0.5	0.79	0.74
0.6	0.87	0.83
0.7	0.93	0.90
0.8	0.96	0.94
0.9	0.98	0.98
1.0	1.00	1.00

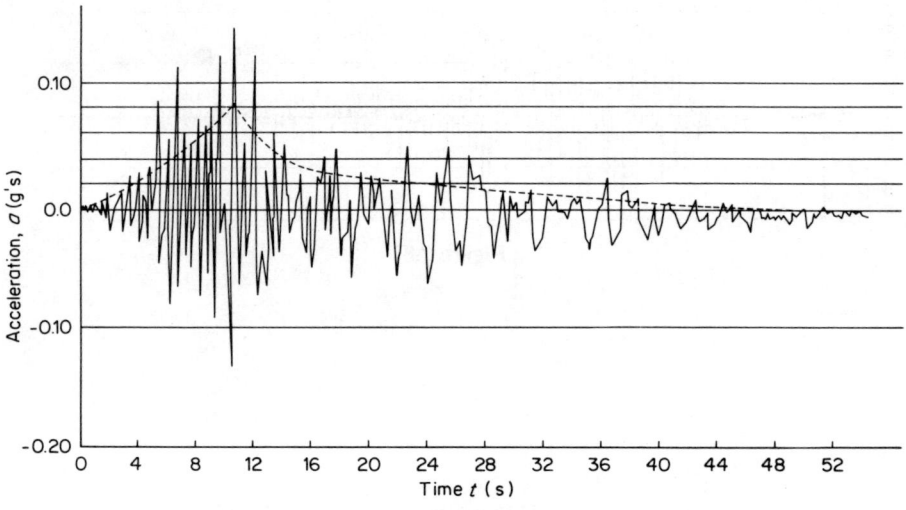

Figure 2.9

Table 2.6 Bucharest, 1977 (see Figure 2.10): $t_f = 14$ s, $\Sigma (a\Delta t)_f = 0.72$, $\cos\theta = 36°$

t/t_f	$\Sigma (a\Delta t)/\Sigma (a\Delta t)_f$	$e^{0.12[1-(t_f/t)^{1.8}]}$
0	0	0
0.1	0.04	~0
0.2	0.19	0.13
0.3	0.46	0.40
0.4	0.63	0.60
0.5	0.73	0.74
0.6	0.81	0.83
0.7	0.88	0.90
0.8	0.93	0.94
0.9	0.97	0.97
1.0	1.00	1.00

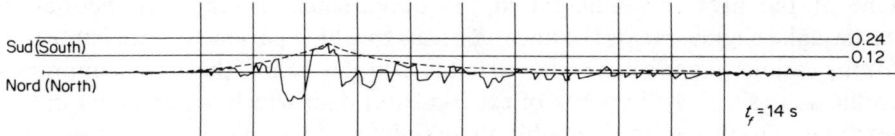

Figure 2.10

3

The Isoseismal Invariant and Its Parameters

INTRODUCTION

One of the problems inherent in the development of rational theories in earthquake engineering is the uncertainty as to which parameters are important for the various phenomena observed in the field. The aim of the present approach to the analysis of the two sets of experimental data which are included in this textbook is to recognize, if possible, these significant parameters. In all cases, this required initial hypotheses or postulates which then had to be checked out against actual observed phenomena.

The basic parameters could be particular physical quantities or they could be invariant models or equations. Both sets are considered and utilized in this text. The previous chapter presented the results of this approach when applied to the accelerogram, which is one of the two sets of field data. This chapter discusses the procedure as it is applied to the second set of experimental data — the isoseismal intensity contour chart. Data from twenty-eight different earthquakes are used to check the basic postulates and the derived expressions.

The intensity contour scale is generally given in descriptive terms, indicating damage or the effects on humans corresponding to different intensities of the earthquakes. The most common intensity scale is the Modified Mercalli Scale (MM) going from 1 to 12 as indicated in the following, reprinted with permission from *Elementary Seismology*, by Charles F. Richter, W. H. Freeman and Company, San Francisco. Copyright © 1958. The scale is also shown in Appendix 2 of Newmark and Rowenblueth.[1] Most intensity scales are quite similar and will be assumed as equivalent in this book.

> To eliminate many verbal repetitions in the original scale, the following convention has been adopted. Each effect is named at that level of intensity at which it first appears frequently and characteristically. Each effect may be found less strongly, or in fewer instances, at the next lower grade of intensity; more strongly or more often at the next higher grade. A few effects are named at two successive levels to indicate a more gradual increase.

Masonry A, B, C, D. To avoid ambiguity of language, the quality of masonry, brick or otherwise, is specified by the following lettering (which has no connection with the conventional Class A, B, C construction).

Masonry A. Good workmanship, mortar, and design; reinforced, especially laterally, and bound together by using steel, concrete, etc.; designed to resist lateral forces.

Masonry B. Good workmanship and mortar; reinforced, but not designed in detail to resist lateral forces.

Masonry C. Ordinary workmanship and mortar; no extreme weaknesses like failing to tie in at corners, but neigher reinforced nor designed against horizontal forces.

Masonry D. Weak materials, such as adobe; poor mortar; low standards of workmanship; weak horizontally.

Modified Mercalli Intensity Scale of 1931 (Abridged and rewritten†)

I. Not felt. Marginal and long-period effects of large earthquakes (for details see text).

II. Felt by persons at rest, on upper floors, or favorably placed.

III. Felt indoors. Hanging objects swing. Vibration like passing of light trucks. Duration estimated. May not be recognized as an earthquake.

IV. Hanging objects swing. Vibration like passing of heavy trucks; or sensation of a jolt like a heavy ball striking the walls. Standing motor cars rock. Windows, dishes, doors rattle. Glasses clink. Crockery clashes. In the upper range of IV wooden walls and frame creak.

V. Felt outdoors; direction estimated. Sleepers wakened. Liquids disturbed, some spilled. Small unstable objects displaced or upset. Doors swing, close, open. Shutters, pictures move. Pendulum clocks stop, start, change rate.

VI. Felt by all. Many frightened and run outdoors. Persons walk unsteadily. Windows, dishes, glassare broken. Knick knacks, books, etc., off shelves. Pictures off walls. Furniture moved or overturned. Weak plaster and masonry D cracked. Small bells ring (church, school). Trees, bushes shaken (visibly, or heard to rustle — CFR).

VII. Difficult to stand. Noticed by drivers of motor cars. Hanging objects quiver. Furniture broken. Damage to masonry D, including cracks. Weak chimneys broken at roof line. Fall of plaster, loose bricks, stones, tiles, cornices (also unbraced parapets and architectural ornaments — CFR). Some cracks in masonry C. Waves on ponds; water turbid with mud. Small slides and caving in along sand or gravel banks. Large bells ring. Concrete irrigation diteches damaged.

VII. Steering of motor cars affected. Damage to masonry C; partial collapse. Some damage to masonry B; none to masonry A. Fall of stucco and some masonry walls. Twisting, fall of chimneys, factory stacks, monuments, towers, elevated tanks. Frame houses moved on foundations if not bolted down; loose panel walls thrown out. Decayed piling broken off. Branches broken from trees. Changes in flow or temperature of springs and wells. Cracks in wet ground and on steep slopes.

IX. General panic. Masonry D destroyed; masonry C heavily damaged, sometimes with complete collapse; masonry B seriously damaged. (General damage to foundations — CFR). Frame structures, if not bolted, shifted off foundations. Frames racked. Serious damage to reservoirs. Underground pipes broken. Conspicuous cracks in ground. In alluviated areas sand and mud ejected, earthquake fountains, sand craters.

X. Most masonry and frame structures destroyed with their foundations. Some well-built wooden structures and bridges destroyed. Serious damage to dams, dikes,

† The author takes full responsibility for this version, which, he believes, conforms closely to the original intention. He requests that, should it be necessary to specify it explicitly, the reference be 'Modified Mercalli Scale, 1956 version,' without attaching his name. The expression 'Richter scale' is popularly attached to the magnitude scale; this is desirable to forestall confusion between magnitude and intensity.

embankments. Large landslides. Water thrown on banks of canals, rivers, lakes, etc. Sand and mud shifted horizontally on beaches and flat land. Rails bent slightly.
XI. Rails bent greatly. Underground pipelines completely out of service.
XII. Damage nearly total. Large rock masses displaced. Lines of sight and level distorted. Objects thrown into the air.

A further basic premise in our analysis of the separate (but related) research areas being discussed was that the key element in all phases of the earthquake engineering happening is the 'energy'. The different topics consider energy from the alternate points of view relevant for the particular subject involved. Thus,

1. Chapter 1 is a study of a new earthquake mechanism in which a closed form non-unique solution is obtained for a possible earthquake-producing occurrence. An approximate value of the energy developed by this mechanism is calculable. The mechanism is consistent with a horizontal fault failure, a type of action which has recently been suggested as a possible deep focus mechanism.
2. Chapter 2 represents a hypothesized theory founded on the accelerogram record for the timewise distribution of energy created by an earthquake mechanism.
3. The present chapter discusses a postulated theory of spacewise distribution of energy from the earthquake, based upon the isoseismal record.
4. Chapter 4 describes a series of charts, figures, and tables based upon Chapters 2 and 3 which are utilized in rational studies of damage assessment (Chapter 5) and also in the direct structural design process (Chapter 7).

NOMENCLATURE FOR THIS CHAPTER

I = Modified Mercalli Scale intensity
S = distance from epicentre to centre of constant intensity region at surface of earth
k, n = non-dimensional constants
f = subscript for final
A, B = numerical constants
H = the total SHE, surface horizontal energy, generated by the earthquake between the epicentre and the radius S_f
Y = derivative with respect to the radius S of the SHE

THE MATHEMATICAL THEORY

An important consideration involved in this text is the attempt to determine the significant parameters connected with rational earthquake engineering analyses. This chapter is concerned with the isoseisms or the isoseismal map. The isoseismal map (see Figure 3.1 which shows the isoseisms of the earthquake which occurred in the Udine region of Italy on June 29, 1873) is a map of the region affected by an earthquake, in which areas of equal intensity are shown. (An

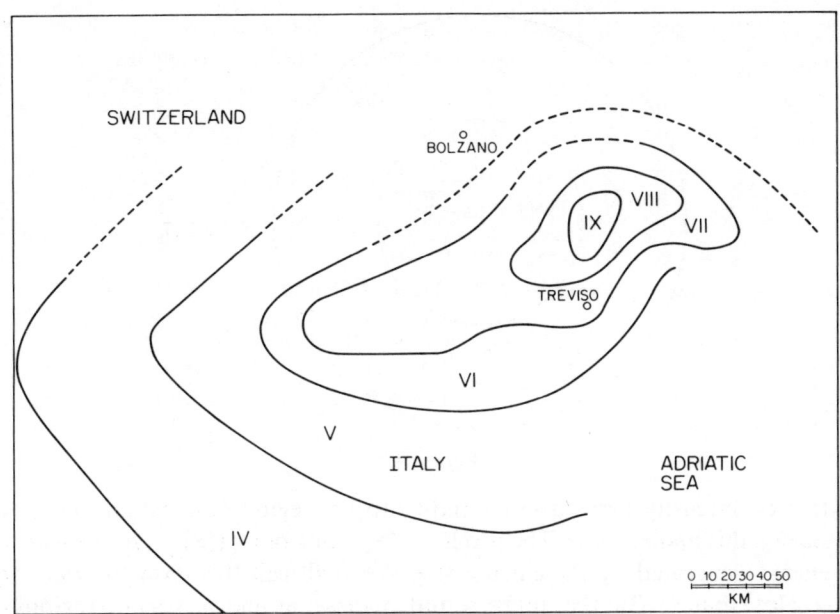

Figure 3.1 Isoseisms of the earthquake occurring on June 29, 1873 in the Udine region of Italy

interesting (and occasionally ignored) aspect of the intensity map is mentioned by Espinosa et al.,[2] wherein they emphasize the rather arbitrary and selective (and frequently incorrect) nature of the intensity contours which are used for many earthquakes. Their comments must surely be appropriate for an earthquake in 1873 shown in Figure 3.1.)

There is no doubt that the location of the regions of different intensity is somewhat arbitrary and subject to the data-takers interpretations. However, it is a recognized set of earthquake engineering data, and an extremely important one. Thus, it would be very desirable that some generalization, or invariant, or parameter, or correlation be developed from this data and the present chapter addresses this problem.

On purely physical grounds, it would seem reasonable that a measure of the earthquake strength — or of its energy — is connected with the intensity number and the area over which this intensity number acts. Thus, as a first step in the development of a model, it was assumed that the isoseismal curves may be represented by equivalent circles. Admittedly, some of the isoseism representations are highly assymetric. These are due to linear surface or near-surface faults. Deep-focus earthquakes probably generate more nearly circular isoseisms. In any case, an average circular representation can be taken as a first step and its validity checked against the realities of actual earthquakes, including the undeniable discrepancies and differences inherent in the determination of accurate isoseismals. Having made the initial 'circular assumption', it is then assumed that a basic parameter for the isoseismal analysis (Figure 3.2) is the

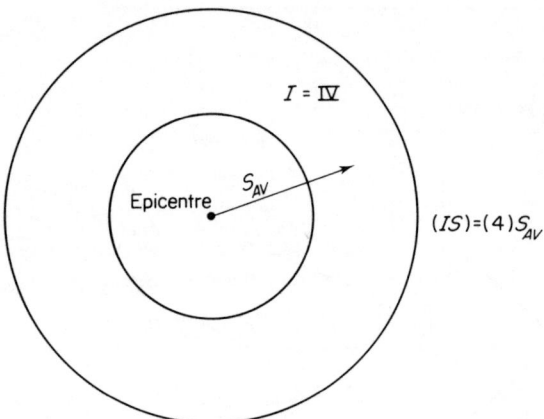

Figure 3.2

product of 'intensity times average radius to the region of constant intensity'. Physically, this makes sense. The product (IS_{AV}) or simply (IS) is one measure of the energy generated by the earthquake. We shall call this term the *Intensity Index*. Note that $S = 0$ at the epicentre and increases as one moves into regions of smaller intensity.

The mathematical development which follows also is reasonable, on mathematical as well as physical bases. We assume

$$\sum_{S=0}^{S=S_{j+1}} (IS) = \sum_{S=0}^{S=S_j} (IS) + \Delta(IS) \Big|_j^{j+1} \tag{1}$$

The next step is a critical one in the mathematical formulation. We assume (note that this preserves dimensional homogeneity and also is physically reasonable),

$$\Delta(IS) = k \sum_{S=S_j} (IS) \frac{S_f^n}{S^{n+1}} \Delta S \tag{2}$$

in which k and n are non-dimensional terms assumed constant. For clarity, the terms in Eqs. (1) and (2) are defined herewith:

I = Modified Mercalli Scale intensity
S = distance from epicentre to centre of constant intensity region
k, n = non-dimensional terms
f = subscript for 'final'. $I = III$ represents a suitable final value and this will be assumed to be the case in this analysis

Now, substituting Eq. (2) in Eq. (1) and proceeding as in Borg,[3] we obtain finally

$$\frac{\sum_{S=0}^{S=S_j} (IS)}{\sum_{S=0}^{S=S_f} (IS)_f} = e^{k[I - (S_f/S)^n]} \tag{3}$$

This equation must be tested against the realities of actual isoseismals.

Eq. 3 is the postulated 'invariant' of the isoseismal representation. It presents a correlation between the Intensity Index and distance from the epicentre of the earthquake. To test this hypothesis, the published isoseismals of twenty-eight different earthquakes were utilized. These earthquakes occurred all over the world during the past 500 years. It was found that the best fit for the data from these earthquakes is given by, approximately,

$$\frac{\sum_{S=0}^{S=S_j}(IS)}{\sum_{S=0}^{S=S_f}(IS)_f} = e^{2[1-(S_f/S)^{1/3}]} \tag{4}$$

At this time it will be assumed that this equation applies to those earthquakes of magnitude equal to or greater than five.

For simplicity, Eq. 4 and the results of five earthquakes only are shown on Figure 3.3. The five earthquakes (which are quite typical) are:

(a) San Fernando, February 9, 1971[4]
(b) Friuli, May 6, 1976[5]
(c) Udine region, June 29, 1873[6]
(d) Washington State, December 14, 1872[7]
(e) Imperial Valley, May 18, 1940[8]

and, as indicated in the Appendix to this chapter, all of the twenty-eight earthquakes satisfy the invariant relation, Eq. 4, satisfactorily.

ANALYSIS OF THE INVARIANT — EQUATION 4

As is clear from Figure 3.3, a plot of the invariant, Eq. 4 is a typical S or growth curve. The point A, which is the point of zero curvature corresponds, physically,

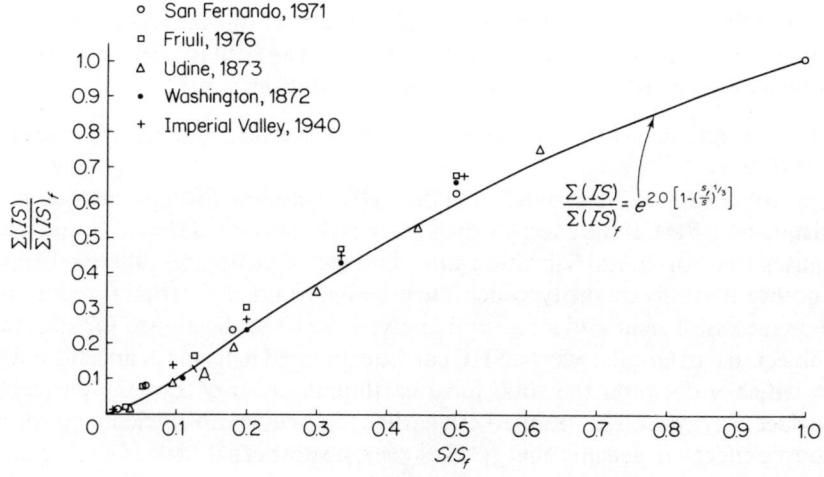

Figure 3.3

to the point at which the rate of change of the sum of the Intensity Index with distance changes from positive to negative. The full importance of this point is not clear at this time. In any case, it occurs at approximately

$$\frac{S}{S_f} = 0.125 \qquad (5)$$

and therefore, from Eq. 4 at the point A

$$\frac{\Sigma(IS)}{\Sigma(IS)_f} = 0.135 \qquad (6)$$

At this time the values given in Eqs. 4, 5, and 6 will be assumed as fixed average values for typical earthquakes, subject to further check, possible change, or modification as required by a more detailed study of isoseisms for earthquakes throughout the world.

THE PHYSICAL VARIABLES AND THE INVARIANT — EQUATION 4

In Eq. 3, S_f and $\Sigma(IS)_f$ vary for each earthquake, k and n are, in fact, constants. The significance of these two statements is related to a tentative uniqueness–existence hypothesis which was formulated by Borg,[9] and which will not be discussed in detail in this book, although the conclusions which follow from it were utilized in Chapter 2 and will now be applied to the isoseismal invariant.

It follows from the hypothesis that there are two fundamental parameters involved in the isoseismal chart analysis. By inspection, it is clear that these are S_f and $\Sigma(IS)_f$, and that these terms vary for different earthquakes. In addition, the tentative theory assumes that this variation of the basic parameters depends upon two physical variables directly related to the isoseismal chart. An analysis of the phenomenon suggests that the two variable quantities are:

1. The average of soil or geological conditions between the epicentre and the distance S_f.
2. The total surface horizontal energy, SHE, between the epicentre and the distance S_f. SHE is the energy which reaches the surface of the earth and which causes the horizontal vibration and damage of structures. In one form or another it is this quantity which must be determined if rational methods of damage assessment and structural analysis are to be developed for structures subject to earthquake effects. SHE can be expressed in terms of an 'efficiency, η' in which η converts the total focal earthquake energy into the destructive surface energy. Bolt, as pointed out earlier,[10] discusses an efficiency to convert source energy to seismic energy. However, he notes that little is known about the value of η.

Furthermore, and once more based upon physical arguments, we shall assume (subject to further verification) that

1a. S_f is the factor influenced by soil or geological conditions.
2a. $\Sigma(IS)_f$ is the factor influenced by the total (SHE) of the earthquake.

The form of the parameters strongly suggests a cross-correlation. Among the indicated topics for future research which this analysis suggests are studies leading either to a confirmation or rejection of the two major correlations hypothesized above.

In any case, the desired invariant, Eq. 4, and the two postulated relations presented above, imply that S_f and $\Sigma(IS)_f$ vary and that these vary depending primarily upon the geological conditions of the region affected by the earthquake and the total SHE of the earthquake.

THE SPACEWISE VARIATION OF SHE

It is possible to develop an expression for SHE per unit radius at a point on the earth's surface where the Intensity Index is $\Sigma(IS)$. This is done by considering a parallel (although different) physically-based mathematical relation for the variation of the $\Sigma(IS)$ term. The procedure is as follows. Assume

$$\frac{\Delta[\Sigma(IS)]}{\Delta S} = A \frac{Y}{H} \Sigma(IS) \qquad (7)$$

in which

    A is a numerical constant
    H is a physical constant, related to the isoseismal map, say 'the total SHE generated by the earthquake between the epicentre and the radius S_f'.

and, based upon dimensional as well as physical requirements,

    Y is the derivative with respect to the radius S, of the SHE.

Now comparing Eq. 7 above and Eq. 2 and equating equivalent terms, we find

$$Y = BH \frac{S_f^{1/3}}{S^{4/3}} \qquad (8)$$

in which B is a constant. Note that Eq. 8 has a singularity at $S=0$, i.e. at the epicentre. The equation will be analysed further in Chapter 4 and will be used in that chapter, and in later chapters, in the damage assessment and structural analysis sections of the textbook. See the footnote on p. 21 for the similar accelerogram–time formulation.

CONCLUSION

The second fundamental source of earthquake experimental data, the isoseismal chart, is analysed and an invariant for this phenomenon is obtained. Two basic parameters are also determined in connection with this analysis, these being the distance S_f and Intensity Index $\Sigma\,(IS)_f$. Various properties of the invariant are obtained and it is shown that — subject to the accuracy of the field data — twenty-eight earthquakes throughout the world over the past 500 years conform to the derived relation. The parameters are also related to the energy of the earthquake and the results obtained in this chapter will be utilized in later chapters dealing with the damage assessment and structural analysis problems in earthquake engineering.

One interesting property which obtains from the analysis of this chapter follows from the fact that Eq. 4 appears to be valid for all earthquakes, even those that are highly assymetric, providing 'average distance' isoseismals are used. Because of this, approximately a given amount of energy is available for each intensity band whether this be circular or a highly asymmetrical shape represented by an average distance ring. This, if verified, would be an important property of earthquakes.

A second interesting fact emerges because Eqs. (5) and (6), also appear to be valid for all earthquakes and imply that for $S/S_f \leqslant 0.125$, the rate of change of Intensity Index sum per unit radius is increasing, whereas for $S/S_f \geqslant 0.125$, the rate of change of Intensity Index sum per unit radius is decreasing.

REFERENCES

1. Nathan M. Newmark and Emilio Rosenblueth. *Fundamentals of Earthquake Engineering*, Prentice-Hall, Inc., Englewood Cliffs, N.J. 1971.
2. A. F. Espinosa, Jose Austurias and A. Quesada. *Applying the Lessons Learned in the 1976 Guatemala Earthquake to Earthquake-Hazard-Zoning Problems in Guatemala*, International Symposium on the Feb. 4, 1976 Guatemalan Earthquake and the Reconstruction Process.
3. S. F. Borg. *Similarity Solutions in the Engineering, Physical–Chemical, Biological–Medical and Social Sciences*, Proceedings of Symposium 'Symmetry, Similarity and Group Theoretic Methods in Mechanics', P. G. Glockner and M. E. Singh (Eds.), Univ. of Calgary, August 1974.
4. Karl V. Steinbrugge *et al. San Fernando Earthquake, Feb. 9, 1971*, Pacific Fire Rating Bureau, 465 California St., San Francisco, Calif. 1971.
5. F. Giorgetti. Isoseismal map of the May 6, 1976 Friuli earthquake, *Boll. Geof. Teor. Appl.*, **XIX**, No. 72, Fig. 1, Udine, Dec. 4–5, 1976.
6. Proceedings of the Specialist Meeting on *The Friuli Earthquake and the Antiseismic Design of Nuclear Installations*, **1**, 114, OECD–NEA/CSNI Rept. No. 28, Rome, Italy, 11–13 October, 1977.
7. Neville C. Donovan. *Let's Be Mean*, NSF Seminar–Workshop on Strong Ground Motion, San Diego, Feb. 1978.
8. Frank Neumann. *Earthquake Intensity and Related Ground Motion*, Univ. of Wash. Press, 1954, as reproduced in Giorgetti,[5] p. 219, Fig. 7.1.
9. S. F. Borg. *Generalized Tensors and Matrices*, Proc. Second International Conference on Mathematical Modeling, St. Louis, Missouri, July, 1979.

10. Robert L. Wiegel (Ed. *Earthquake Engineering*, Chap. 2 written by Bruce A. Bolt, Prentice-Hall, Inc., Englewood Cliffs, N.J., 1970.
11. S. F. Borg. *An Isoseismal–Energy Correlation for Use in Earthquake Structural Design*, Seventh World Conf. on Earthquake Engineering, Istanbul, Turkey, Sept. 8–13, 1980. Also Technical Report ME/CE–792 Dep't of Mechanical Engineering (Civil Engineering), Stevens Institute of Technology, Hoboken, N.J., Dec. 1979.
12. F. Machado. *Seismology-Portugal: Curso de Sismologia*, Imprimarte SARL, Losboa 1970.
13. National Academy of Sciences. *Seismology, Responsibilities and Requirements of a Growing Science*, Part II, 1970.
14. M. Bath. *Introduction to Seismology*, Berkhauser Verlag, Basil, Borton, Stuttgart, Germany, 1979.
15. K. E. Bullen. *New Zealand Seismology*, Methuen, London, 1954.
16. Australian Congress. *Earthquake Engineering Symposium*, Melbourne, 1969.
17. E. J. Ramirez. *Historia de los Terremotos en Colombia*, Institute Geografico Agustin Codazzi, Bogota, 1969.
18. AISI. *The Agadir Morocco Earthquake*, American Iron and Steel Institute, N.Y. 1962.
19. Indonesia Earthquakes. *Lembaga Meteorolog dan Geofisika*, Ministry of Communications, Meteorological and Geophysical Institute, Jakarta, 1971.
20. G. L. Berlin. *Earthquakes and the Urban Environment*, Vol. II, CRC Press Inc., West Palm Beach, Fla. 1978.
21. James M. Gere and Haresh C. Shah. *Tangshan rebulds after mammoth earthquake*, Dec. 1980, Civil Engineering.
22. The New York Times, Fyn. 16, 1981.
23. S. F. Borg. *Extended Analysis of Isoseismal–Magnitude–Intensity Index Correlations in Earthquake Engineering*, Technical Report ME/CE–81-1, Department of Mechanical Engineering (Civil Engineering), Stevens Institute of Technology, Hoboken, N.J., May, 1981.

APPENDIX

Eqs. (4) and Figure 3.3 were checked against twenty-eight different earthquakes in various parts of the world that had occurred in the past five hundred years. Detailed computations are shown for four of these. References and a summary of all twenty-eight are shown in the Summary Table 3.5.

The model was brought down to intensity III, which agrees with the other checks made and appears to be about the smallest intensity that can usually be observed or felt. Since the map[5] only went down to intensity VII, the lower intensity values were obtained from a line drawn on the Neumann curve[8] parallel to the lines previously drawn for the California earthquakes. This appears to be reasonable since the higher intensity values (obtained from Figure 3.3) do fall on this assumed line.

A final statement concerning Figure 3.3. It is clearly possible to obtain a closer fit of the data by choosing other values for k and n. Also, k and n are not unique (although there is little variation possible in these for a close fit). It is not deemed worthwhile to belabour the point concerning k–n values. It must be remembered that the data itself — the isoseismal maps — are extremely subjective and representations of the same event by different investigations would result in more variation than shown in Figure 3.3. Eq. 4 may be pointed to as just one of a

Table 3.1 San Fernando, February 9, 1971[4,11]

I	Distance to outer reach (miles)	S_{AV}	(IS)	$\dfrac{S_f}{S}$	$\dfrac{S}{S_f}$	$\dfrac{\Sigma(IS)}{\Sigma(IS)_f}$
Epicentre	0	0.	0	∞	0	0
VIII–IX	5	2.5	21.3	72	0.014	0.018
VII	15	10	70	18	0.06	0.08
VI	50	32.5	195	5.5	0.18	0.24
V	130	90	450	2	0.50	0.62
I–IV (SAY. II 1/2)	250	190	450	1	1.0	1.00

Table 3.2 Friuli, May 6, 1976[5,11]

I	Distance to outer reach (km)	S_{AV}	(IS)	$\dfrac{S_f}{S}$	$\dfrac{S}{S_f}$	$\dfrac{\Sigma(IS)}{\Sigma(IS)_f}$
Epicentre	0	0	0	∞	0	0
X	8	4	40	125	0.01	0.008
IX	17	12.5	112	40	0.02	0.032
VIII	38	27.5	220	18	0.06	0.08
VII	85	62	434	8	0.12	0.17
VI	120	102	612	5	0.20	0.30
V	205	162	810	3	0.33	0.47
IV	300	252	1008	2	0.50	0.68
III	700	500	1500	1	1.0	1.00

Table 3.3 Udine Region, June 29, 1973[6,11]

I	Distance to outer reach (km)	S_{AV}	(IS)	$\dfrac{S_f}{S}$	$\dfrac{S}{S_f}$	$\dfrac{\Sigma(IS)}{\Sigma(IS)_f}$
X	0	0	0	∞	0	0
IX	12	6	54	34	0.03	0.02
VIII	27	19	152	10.8	0.09	0.09
VII	45	36	252	5.7	0.18	0.19
VI	80	63	378	3.3	0.30	0.35
V	100	90	450	2.3	0.43	0.53
IV	160	130	520	1.6	0.63	0.75
III	250	205	615	1.0	1.00	1.00

Table 3.4 Imperial Valley, May 18, 1940[8,11]

I	Distance to outer reach (km)	S_{AV}	(IS)	$\dfrac{S_f}{S}$	$\dfrac{S}{S_f}$	$\dfrac{\Sigma(IS)}{\Sigma(IS)_f}$
X	0	0	0	∞	0	0
IX	3	1.5	14	50	0.02	0.02
VIII	5	4	32	18.8	0.05	0.07
VII	9	7	49	10.7	0.09	0.14
VI	20	15	90	5	0.20	0.27
V	30	25	125	3	0.30	0.45
IV	50	40	160	1.9	0.53	0.68
III	100	75	225	1.0	1.00	1.00

Table 3.5 Summary data for twenty-eight earthquakes

Designation	Earthquake	S_f km	$\Sigma(IS)_f$, S in km	Ref.
a	San Fernando, 1971	300	2360	4
b	Friuli, 1976	500	5000	5
c	Udine, 1873	205	2420	6
d	Washington State, 1872	900	8000	7
e	Imperial Valley, 1940	75	700	8
f	Valparaiso, 1906	1100	15000	1
g	Mexico, 1962	390	4700	1
h	S. Miguel, 1522	115	1620	12
i	Lisbon, 1755	1500	23350	12
j	S. Jorge, 1757	125	1970	12
k	Charleston, 1886	1500	23200	13
l	San Fransisco, 1906	550	4420	14
m	Messina, 1908	335	4430	12
n	Benavente, 1909	340	3450	12
o	Hawke's Bay, 1921	440	4380	15
p	Faial, 1926	100	1160	12
q	Wairoa, 1932	250	2600	15
r	Madeira, 1941	1150	14500	12
s	Wewak, 1946	280	2450	16
t	Orleanville, 1954	95	1375	12
u	Arboledos, 1950	80	1040	17
v	Agadir, 1960	30	460	18
w	Madjene, 1969	65	760	19
x	Agadir, 1969	50	725	12
y	Bantarkawung, 1971	75	730	12
z	Lice, 1975	80	780	20
aa	Tangshan, 1976	700	9770	21
bb	Corinth, 1981	95	1040	22

number of different equations that could be used to represent the variation. This is true — but Eq. 4 is useful in several respects. It is dimensionally sound, it has reasonable physical and mathematical bases, and, furthermore, it permitted an extension corresponding to Eq. 8.

It is a hypothesized equation. It must be checked out further. But it appears to be a very promising first step in the direction of developing a rational, analytical knowledge of a particular phase of the earthquake problem and it does lend itself to extended applications to the damage assessment and structural design problems.

In Table 3.5, on page 39, are shown the key values for twenty-eight different earthquakes which have occurred in the past 500 years all over the world. Each of these has an isoseismal map which corresponds to Eq. 4 and Figure 3.3, subject to the probable accuracy of the data as reported in the references. In this Table are shown values for $\Sigma(IS)_f, S_f$ and the references from which the data are obtained. The detailed tabulations for those earthquakes not shown in Tables 3.1 to 3.4 are given in Borg.[23]

4

The Earthquake Engineering Design Charts

INTRODUCTION

As noted previously, the main thrust and purpose of this textbook is to develop earthquake theories and procedures that can be used by engineering design offices and others involved with the damage assessment and structural design aspects of the event. Furthermore, it was pointed out that derived invariants and parameters were to be utilized and these in conjunction with energy, since energy in one form or another is at the core of all phases of the earthquake phenomena.

In the previous two chapters, the accelerogram and the isoseismal chart were analysed from the invariant–parameter point of view and it was shown that these can be represented, to an acceptable degree of engineering accuracy, by means of a suitable invariant in terms of particular parameters which hold for the two basic sets of data.

In this chapter, these results are extended and a series of engineering design charts will be described. These charts, whose parameters are those previously described as well as other well-known variables of earthquake engineering such as magnitude of earthquake and efficiency of earthquake, will be the basic design tools for analysing earthquake damage assessment and structural design. Each chart will be discussed separately (starting, for completeness, with the invariant curves of Chapters 2 and 3). Those to be used for damage assessment will be discussed in more detail in Chapter 5. Those that are to be used for the structural analysis and design will be utilized in Chapter 7.

Some additional comment is in order concerning the accuracy of the two basic sources of data. In the first place, it must be recognized that the accelerogram and isoseismal are two major sources of direct data available for the effect of an earthquake. Some of the uncertainty has been pointed out: 'Following a strong earthquake, different stations report magnitudes often differing more than one degree from each other.' Within 48 hrs these values are unified to within 0.2 degrees, giving the impression of great precision. The spread in intensities assessed by different observers is even greater'.[1]

The charts and curves developed in this and earlier chapters are not 'exact' (even if we assume that exactness is possible in any engineering analysis).

However, it must also be recognized that the theories and developments presented do have sound physical, mathematical, and dimensional bases. In their entirety, they form a rational, interrelated, coherent theory of the earthquake response, suitable for engineering design and related purposes (such as damage assessment) by the average engineering office. The charts are self-correcting, that is, in their present, primitive form they are subject to revision and greater precision should follow as more data is collected and correlated with derived expressions.

NOTATION FOR THIS CHAPTER

For convenience, the different terms used in this chapter are collected and presented below:

a = ordinate to accelerogram envelope
f = subscript, final
i = subscript, small lower value
t = time
B = a numerical constant
D = distance
E = energy, in ergs, of earthquake
G = a numerical constant
H = total surface horizontal energy of the earthquake between the radius S_i and up to and including the Intensity = III radius, S_f
I = Intensity of the earthquake
K = a constant
M = magnitude of the earthquake
$\mathcal{R}$ = designation for geological region
S = distance from epicentre to mid-radius of an intensity band
T = derivative of the surface horizontal energy with respect to the time, t, due to the earthquake at the accelerogram location
W = total surface horizontal energy between the small inner radius, S_i, and the radius S
Y = derivative of the surface horizontal energy with respect to the radial distance S
ε_t = surface horizontal energy per unit area between the times t_i and t at the accelerogram location
θ = angle made by accelerograph direction and a ray from the epicentre
η = efficiency of the earthquake, i.e. $H = \eta E$

THE ACCELEROGRAM CHART AND PARAMETERS

As noted in Chapter 2, a canonical invariant can be derived for the accelerogram in terms of a bounding envelope and the two fundamental parameters $\Sigma(a\,\Delta t)$, the Acceleration Index, and t for this canonical accelerogram. The resulting

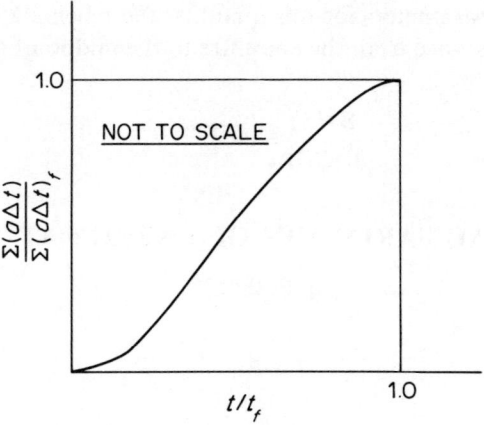

Figure 4.1

invariant is given by Eq. 1 and Figure 4.1 which follow (these have been reproduced from Chapter 2).

$$\frac{\Sigma\,(a\,\Delta t)}{\Sigma\,(a\,\Delta t)_f} = e^{0.12[1-(t_f/t)^{1.8}]} \tag{1}$$

THE ISOSEISMAL CHART AND ITS PARAMETERS

As indicated in Chapter 3, if the 'circular approximation' assumption is made, then isoseismal charts for approximately $M \geqslant 5$ earthquakes can be represented by the following equation and the curve in Figure 4.2, these being the invariant for this very important source of field data. Note, Eq. 2 and Figure 4.2 are given in

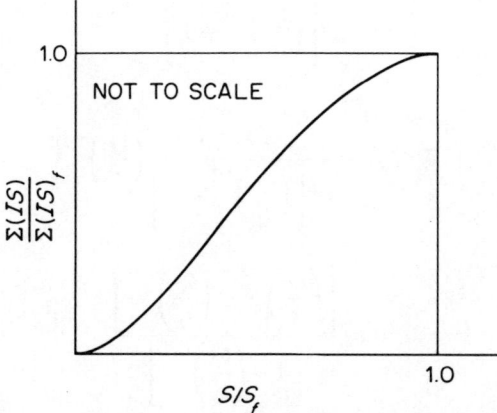

Figure 4.2

terms of the basic parameters for this quantity, these being $\Sigma(IS)$, the Intensity Index and S, the distance from the epicentre to the midpoint of the $I=$ constant intensity band.

$$\frac{\Sigma(IS)}{\Sigma(IS)_f} = e^{2.0[1-(S_f/S)^{1/3}]} \tag{2}$$

THE TEMPORAL VARIATION OF ENERGY AT A POINT

In Chapter 2, it was shown in Eq. 9, that

$$T = B\varepsilon_{t_f} \frac{t_f^{1.8}}{t^{2.8}} \tag{3}$$

in which

T = derivative of the surface horizontal energy with respect to the time, t, due to the earthquake at the accelerogram location
B = a numerical constant
ε_{t_f} = the total surface horizontal energy per unit area supplied by the earthquake as the point where the accelerogram is obtained
t, t_f = any time t and the total time, t_f, of the accelerogram record

Using Eq. 3 we can determine the value of B by integrating over the interval $\int_{t_i \to 0}^{t_f}$, in which t_i is a very small time, and it is introduced in order to surmount the infinity at $t=0$. Then

$$\int_{t_i}^{t_f} B\varepsilon_{t_f} \frac{t_f^{1.8}}{t^{2.8}} dt = \int_{t_i}^{t_f} T \, dt = \varepsilon_{t_f} \tag{4}$$

and therefore

$$T = \frac{\varepsilon_{t_f}}{1.8\left[\left(\frac{t_f}{t_i}\right)^{1.8} - 1\right]} \frac{t_f^{1.8}}{t^{2.8}} \tag{5}$$

also

$$\int_{t_i}^{t} T \, dt = -\frac{1.8\varepsilon_{t_f}}{1.8\left[\left(\frac{t_f}{t_i}\right)^{1.8} - 1\right]} \left(\frac{t_f}{t}\right)^{1.8}\bigg|_{t_i}^{t}$$

$$= \varepsilon_{t_f} \frac{\left[\left(\frac{t_f}{t}\right)^{1.8} - \left(\frac{t_f}{t_i}\right)^{1.8}\right]}{\left[1 - \left(\frac{t_f}{t_i}\right)^{1.8}\right]} \tag{6}$$

$$= \varepsilon_t$$

where ε_t = surface horizontal energy per unit area between the times t_i and t at the accelerogram location.

Thus finally

$$\frac{\varepsilon_t}{\varepsilon_{t_f}} = \frac{\left[\left(\frac{t_f}{t_i}\right)^{1.8} - \left(\frac{t_f}{t}\right)^{1.8}\right]}{\left[\left(\frac{t_f}{t_i}\right)^{1.8} - 1\right]} \quad (7)$$

and clearly as t goes from $t_i \to t_f$, this ratio goes from $0 \to 1$ as required.

The question which must be answered now is — 'What is the value of t_i?' An inspection of typical canonical accelerograms indicates that if we take

$$\frac{t_f}{t_i} \cong 10 \quad (8)$$

the results (at least initially) appear to be acceptable for engineering purposes. Subject to revision as more data indicates, we shall use this value, from which, finally, the temporal variation of surface horizontal energy at a point is given by

$$\frac{\varepsilon_t}{\varepsilon_{t_f}} = \frac{63 - \left(\frac{t_f}{t}\right)^{1.8}}{62} \quad (9)$$

which plots as shown on Figure 4.3. Also shown on this figure are curves corresponding to other values of t_i/t_f. Figure 4.3 may be used in design applications, subject to revision as more data is collected.

THE SPACEWISE VARIATION OF ENERGY OVER THE ENTIRE FIELD

In Chapter 3, it was shown that the variation of surface horizontal energy with distance over the area affected by an earthquake is given by

$$Y = BH \frac{S_f^{1/3}}{S^{4/3}} \quad (10)$$

The various terms are defined as follows:

Y = derivative with respect to the radius, S, of the SHE generated by the earthquake
B = a constant
S_i = a small inner radial distance, lower limit of the integration to remove the singularity at $S = 0$
S = radial distance in the earthquake field
S_f = radial distance to the centre of the I = III field, the maximum distance in this analysis

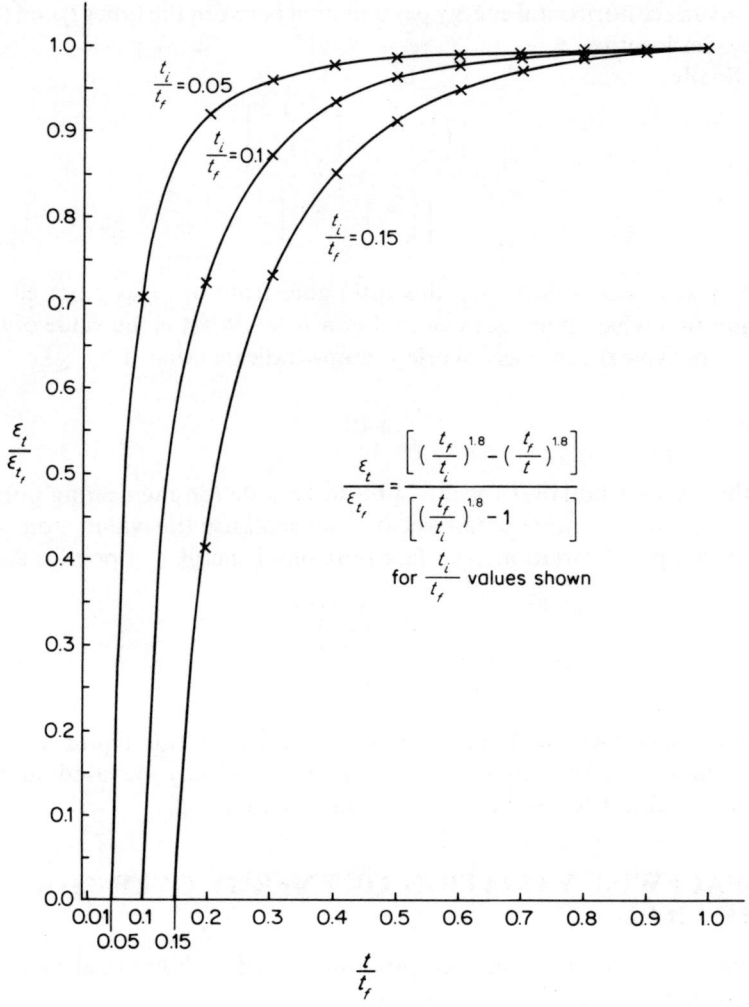

Figure 4.3

H = total surface horizontal energy of the earthquake between S_i and the Intensity = III radius, S_f

W = total surface horizontal energy between the small inner radius S_i and the radius S

If we integrate Eq. 10 between the limits $S_i \to S$ we obtain

$$W = KH\left[\left(\frac{S_f}{S_i}\right)^{1/3} - \left(\frac{S_f}{S}\right)^{1/3}\right] \tag{11}$$

K = a constant which we evaluate by noting that $W = H$ when $S = S_f$, so that

$$K = \frac{1}{\left[\left(\frac{S_f}{S_i}\right)^{1/3} - 1\right]} \qquad (12)$$

and therefore

$$\frac{W}{H} = \frac{\left[\left(\frac{S_f}{S_i}\right)^{1/3} - \left(\frac{S_f}{S}\right)^{1/3}\right]}{\left[\left(\frac{S_f}{S_i}\right)^{1/3} - 1\right]} \qquad (13)$$

Concerning the S values and, in particular, the fact that they have a dual nature, i.e.

(1) in the isoseismal index invariant, Eq. 2, they have descrete values, S_{III}, S_{IV}, etc.
(2) in the energy variation over the field equation, Eq. 13 they have continuous values, i.e. energy $= f(S)$.

This behaviour is not unknown in applied mechanics and physics. The performance of S noted above is similar to (but not the same as) the eigenvalue conduct of, say, the column buckling problem in which a term kx/l has the values $\pi, 2\pi, \ldots n\pi$ for determining the critical buckling loads, P_{cr}, although the bending moment in the member, as a function of x is continuous for $0 \leqslant x \leqslant l$. There is also the particle-field representations in electromagnetic wave theory.

In this earthquake engineering analysis, the analogous procedure is that in (1) above, S has the values S_{III}, S_{IV}, etc., for determining the isoseismal index relation, but the energy throughout the earthquake region as a function of S is continuous for $S_i \leqslant S \leqslant S_f$.

The question arises concerning the value of S_i, the 'small' inner radius. Because of the 1/3 power, the choice is somewhat insensitive to most reasonable assumptions. To indicate one form of the variation, Table 4.1 which follows shows values of W/H for several values of S_f/S_i and a particular value of S_f/S. Considering the uncertainty in the value of H, the numbers shown in the last column represent an acceptable spread.

It may be shown (this will be done in Chapter 7) that W is related to ε_{t_f} of Eq. 3. Also, it is reasonable to assume that ε_{t_f} has some connection with the isoseismal contours and hence with the measure of damage. This, in turn, may permit an approximate determination or calibration of S_f/S_i (using the charts and methods

Table 4.1 Values of W/H

S_f/S_i	$(S_f/S_i)^{1/3}$	(W/H) for $(S_f/S)^{1/3} = 2$
1,000,000	100	0.99
1,000	10	0.89
125	5	0.75

described in the following sections) by determining the effects of various earthquakes on 'standard' structures, pendulums or other apparatus at given distances from the epicentres. (Another, perhaps preferable, procedure which suggests itself is to assume S_i as the radius to the Intensity = IX or X contour and to determine an earthquake efficiency, η, corresponding to this. In this way, the method will be standardized and the energy to be absorbed by the building (which depends upon S_i and η as shown in Chapter 7 and which therefore will assist the designer in his choice of η) can be obtained with greater accuracy as experimental data (i.e. design data) accumulates.)

It is possible that S_f/S_i varies for different geological conditions. This can only be determined following the calibration procedure described above. For a more detailed discussion of this and related points see the statements in Chapter 5.

For preliminary, tentative design purposes, subject to revision, we shall assume that $S_f/S_i = 1000$ or Eq. 13 becomes

$$\frac{W}{H} = \frac{10 - \left(\frac{S_f}{S}\right)^{1/3}}{9} \qquad (14)$$

for $S_i \leqslant S \leqslant S_f$. See also Figure 4.4, a plot of Eq. 14 and also of several other curves corresponding to the S_i/S_f values shown.

THE 'GEOLOGY' OF A REGION

We must assume that the effect of an earthquake on structures is related to the mechanism and also to the depth of focus of the earthquake as well as to the overall geological character of the earthquake field. To attempt an exact analysis including these as variables (in addition to other possible 'geological' factors) would be an impossible task. We can, however, assume *three* separate, different geological regions as an approximation to the effect of geology on the engineering response.

A three-region classification is given[2] in terms of:

1. The Circumpacific Belt,
2. The Alpide Belt,
3. The Low Seismicity Region.

In this chapter and in later sections of the text we shall use a somewhat different scheme with designations as noted:

1. Surface Fault Region, $\mathcal{R}_1$.
2. Mountain Region, $\mathcal{R}_2$.
3. Plains Region, $\mathcal{R}_3$.

In a very rough way, the focal depth increases with the number designation given above and therefore the effect of focal depth, for simplicity, will be assumed

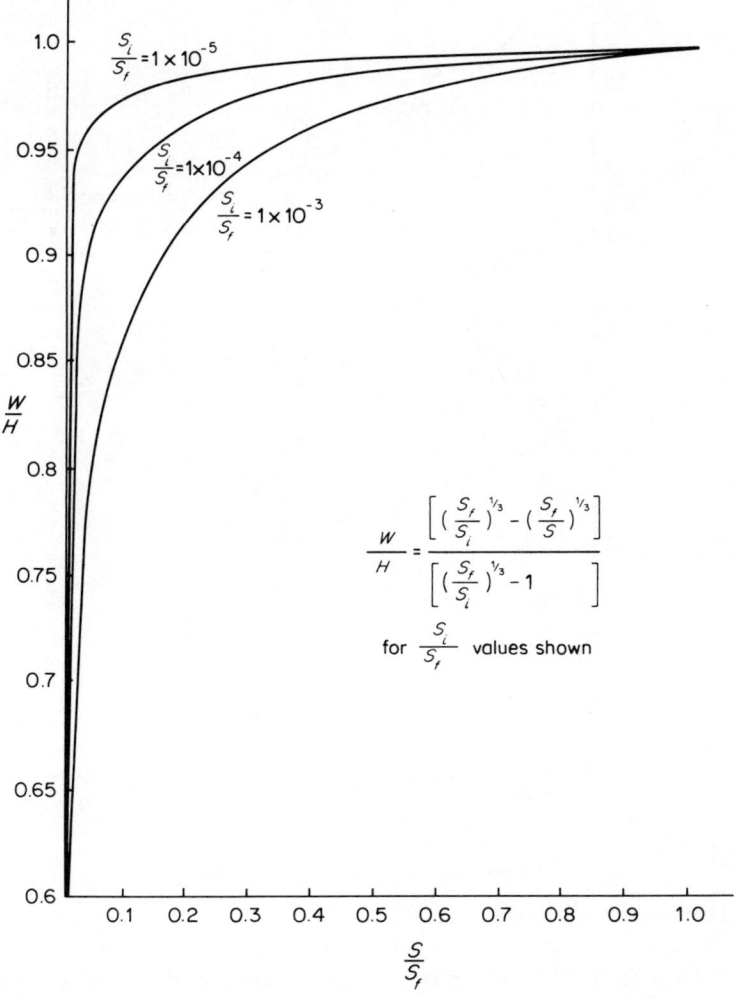

Figure 4.4

as accounted for in the geologic region designation. The frequencies of the accelerations also will be assumed as roughly correlated with the geology. See the discussion of this point in the next section.

INTENSITY AS RELATED TO THE ACCELEROGRAM PARAMETERS AND GEOLOGY

t_f and $\Sigma (a \Delta t)_f$ are assumed to be the fundamental parameters in our formulation of the accelerogram invariant. Furthermore, on physical grounds, it seems reasonable to assume that the damage inflicted on a structure at a given location is related to the acceleration of the ground at the location. It appears reasonable

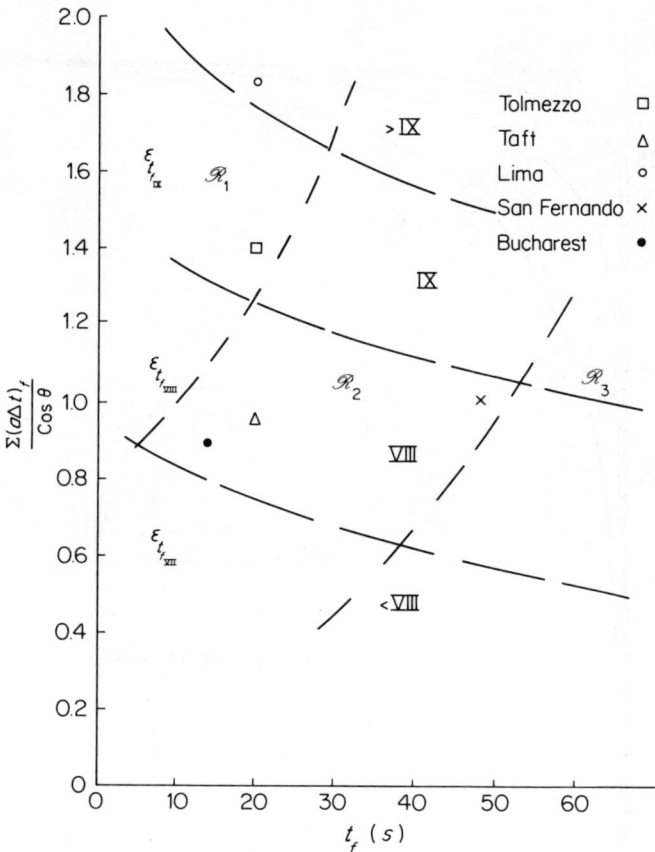

Figure 4.5 Damage contour chart

therefore to assume that the parameters $\Sigma(a\,\Delta t)_f$ and t_f are in some way connected with the intensity of an earthquake at the location of the accelerogram instrument subject to variations due to geology.[3,4]

A preliminary form of this correlation is shown in Figure 4.5, in which the direction of the accelerogram is accounted for, as well as geology, for the five canonical accelerograms considered in Chapter 2. The data for $\Sigma(a\,\Delta t)_f$ and t_f are given on Table 2.1 using Eq. 7, with $\cos\theta$ taken as 1.0 for those cases where it is not known. We have therefore the values as shown in Table 4.2.

Also shown in Figure 4.5 are tentative intensity contours and geology regions, as well as general designations for total energy per unit area as related to intensity. Another factor that almost certainly is involved in damage analyses are the frequencies of the accelerogram. These have not been brought into the model at this point, but it is conceivable that some sort of 'average frequency' influences the structural damage and another subject suitable for study is the possibility of bringing the effect of frequency into the construction of the Damage Contour

Table 4.2 $\Sigma (a \Delta t)_f$ and t_f

Earthquake	t_f, s	$\Sigma (a \Delta t)_f$, 'g' s	θ	$\dfrac{\Sigma (a \Delta t)_f}{\cos \theta}$	Intensity I
Tolmezzo, 1976	20	0.98	45°	1.39	IX
Taft, 1952	20	0.96	unknown	0.96	VIII
Lima, 1966	20	1.84	unknown	1.84	>IX
San Fernando, 1971	48	1.06	unknown	1.06	VIII
Bucharest, 1977	14	0.72	36°	0.89	VIII

Chart, Figure 4.5. Perhaps different charts are required, subject to some suitable measure of frequency and the geology.

Some hints concerning this last point may come from the theory developed in Chapter 7. In that chapter, a method of structural analysis is presented which utilizes the invariants of this book. Using the method described it is possible to consider the effects of different frequencies in the overall behaviour of the structure. Hence a related topic for future study is to determine the effect, on a structure, of typical accelerograms of given energy input, with different frequency patterns. The most critical case could be looked for and this will give some fundamental insight into the structural damage phenomenon.

Clearly more points are needed in order to establish the validity of Figure 4.5 as well as its accuracy. Also the ε_{t_f} energy values must be determined following calibration of the intensity–energy contours. It is also conceivable that additional geological regions — $\mathscr{R}_4$, etc. — are required to represent the relations more accurately.

THE MAGNITUDE–$\Sigma (IS)_f$–S_f CORRELATIONS

Just as the intensity and local energy can be approximately correlated with the local accelerogram parameters, so also we shall assume that the magnitude, M, of the earthquake can be related to the overall field parameters, $\Sigma (IS)_f$ and S_f. To test this hypothesis we consider the data shown in the earthquake chart in Chapter 3 (Table 3.6) in which isoseismals were analysed for many earthquakes all over the world. The chart contains $\Sigma (IS)_f$ and also S_f values. Magnitudes are shown for eight of the earthquakes considered (Table 4.3) and some preliminary information may be obtained from an analysis of the data as suggested above. The data from Table 3.6 are repeated in Table 4.3 for convenience.

Figure 4.6 shows a plot of the data for the 28 earthquakes. Two observations may be made concerning this figure.

1. The points generally fall along the line shown. We may consider this a good correlation when we remember that the data goes back 500 years and recall the comment concerning accuracy in the Introduction to this chapter.

Table 4.3

Designation	Earthquake	$\Sigma (IS)_f$ S in km	M, in ref. when given	S_f, km
a	San Fernando, 1971	2360	6.50	300
b	Friuli, 1976	5000		500
c	Udine, 1873	2420		205
d	Washington State, 1872	8000		900
e	Imperial Valley, 1940	700		75
f	Valparaiso, 1906	15000		1100
g	Mexico, 1962	4700		390
1	S. Miguel, 1522	1620		115
2	Lisbon, 1755	25350		1500
3	S. Jorge, 1757	1970	7.40	125
4	Charleston, 1886	23200		1500
5	San Francisco, 1906	4420	8.20	550
6	Messina, 1908	4430		335
7	Benavente, 1909	3450		340
8	Hawke's Bay, 1921	4380		440
9	Faial, 1926	1160		110
10	Wairoa, 1932	2600		250
11	Madeira, 1941	14500		1150
12	Wewak, 1946	2450		280
13	Orleavnille, 1954	1375	6.75	95
14	Arboledas, 1950	1040		80
15	Agadir, 1960	460	5.75	30
16	Madjene, 1969	760		65
17	Agadir, 1969	725		50
18	Bantarkawung, 1971	730		75
19	Lice, 1975	780	6.70	80
20	Tangshan, 1976	9770	7.8	700
21	Corinth, 1981	1040	6.60	95

2. There appear to be different sets of magnitude groupings based upon the very limited data available. For the eight earthquakes whose magnitudes are shown, it is possible to introduce a tentative geologic correlation corresponding to the three regions presiously described. These are as follows and are shown on Figure 4.6.

$\mathscr{R}_1$	$\mathscr{R}_2$	$\mathscr{R}_3$
San Fernando	Lice	Agidir
San Francisco	S. Jorge	Corinth
		Orleanville
		Tangshan

Some of the $\mathscr{R}_2$ and $\mathscr{R}_3$ earthquakes should perhaps be interchanged. As more data is collected and sorted out in accordance with the classification proposed herein it will be possible to obtain a clearer, more accurate representation of

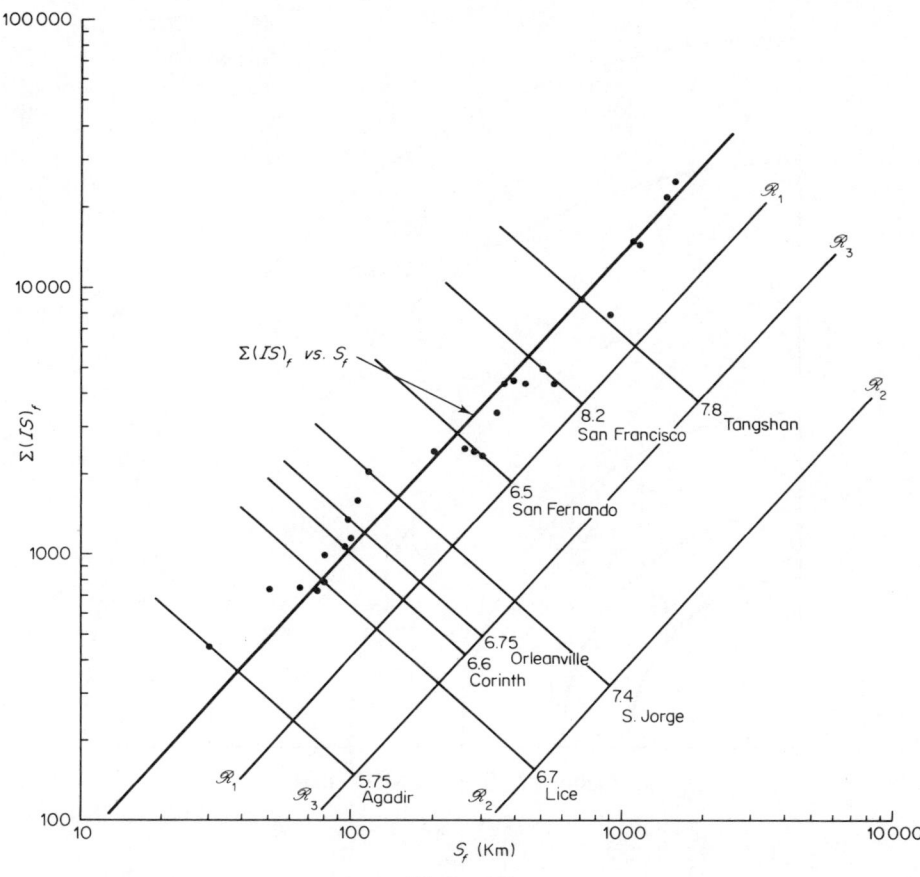

Figure 4.6

Figure 4.6. More data is needed, more points for more earthquakes all over the world, in order to establish the magnitude effect. Even so, it seems reasonable that Figure 4.6 can be used to obtain approximate correlations between earthquake magnitude, M, and $\Sigma(IS)_f$ and S_f values for engineering design purposes. As will be seen in the next section, these will be used to prepare the MID charts, and to assess damage to structures due to earthquakes.

THE MAGNITUDE–INTENSITY–DISTANCE (MID) CHART

One final set of charts that is useful in the damage assessment as well as the structural design–analysis portions of earthquake engineering is the magnitude–intensity–distance (or MID) charts. These curves contain information that will be of value to insurers as well as to engineers, since they will permit one to estimate probable structural damage due to a particular earthquake. In addition architects and preparers of design codes should find use for the charts.

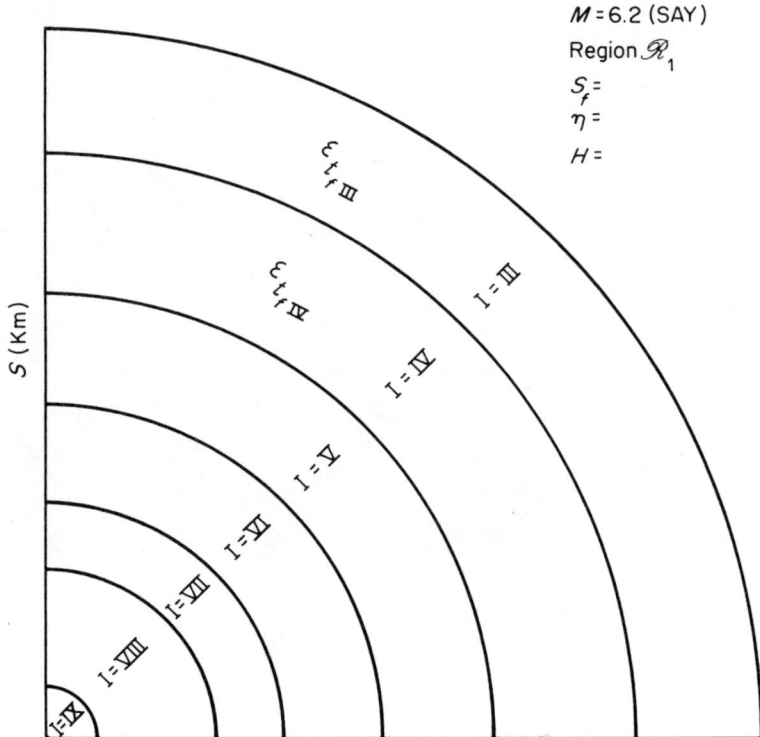

Figure 4.7 Typical MID chart

Typically, the charts will consist of a series of curves similar to the one shown in Figure 4.7. All three geologic regions will be represented with magnitudes going from say 5 to 8.5, in increments of 0.1 or 0.2. Values of $\Sigma\,(IS)_f$ and S_f will be obtained from Figure 4.6 as this chart is continuously revised and refined and made more accurate. In addition, values of W/H as determined from Eq. 14 may be shown on the charts, with H to be determined from the energies corresponding to M (Ref. 1) and assumed values of the earthquake efficiency, η,* a term which converts M energy into surface horizontal energy, H. Knowing W at each S, an approximate εt_f-intensity correlation can be determined.

A particular chart is prepared as follows:

1. Assume a value of M for a given geological region.

---

* An approximate theoretical determination of the efficiency, η, in terms of depth of focus (and therefore, roughly, the geology) and also of the magnitude, M, (for say, $5 \sim M \sim 8$) indicates that the efficiency has values $0 \sim \eta \sim 0.20$. In this evaluation based upon the compatibility of energy requirement for equal — numbered intensity bands, it is assumed that the efficiency varies inversely with both the depth of focus and with the magnitude. Thus, a relatively moderate California earthquake (say $5 \sim M \sim 6$) will tend to cause more overall damage than an earthquake of the same magnitude in, say, South Carolina.

2. Determine the energy E, due to M, Ref. 1

$$\text{Log}_{10} E \cong 11.4 + 1.5M \tag{15}$$

with E in ergs.
3. Assume a value for the efficiency, η.
4. Determine $H = \eta E$.
5. From Figure 6, knowing M and the geology, determine $\Sigma (IS)_f$ and S_f.
6. From Eq. 2, with known $\Sigma (IS)_f$ and S_f, start with Intensity = III and S_f, so that

$$\frac{\Sigma (IS)_f = 3S_f}{\Sigma (IS)_f} = e^{2[1-(S_f/S_{IV})^{1/3}]} \tag{16}$$

and determine S_{IV} the only unknown in this equation.
7. Repeat step 6 for $S_V, S_{VI} \ldots S_X$, reducing the numerator of the left hand side by, successively, $4S_{IV}, 5S_V \ldots 9S_{IX}$.
8. Having S and S_f for the entire field, from Eq. 14 determine W/H and from step 4, finally, W. If (and when) Eq. 14 is modified as more and more calibrating data becomes available, then W will, of course, be changed accordingly.
9. Knowing W at any S, it is possible to determine an approximate ε_{t_f} at each S.

CONCLUSION

As noted in earlier sections of the textbook, a basic purpose of this presentation has been to determine the invariants and parameters of the earthquake event. Toward this end the two main observables, or sources of experimental data, were to be used. These are

1. The accelerogram.
2. The isoseismal map.

The invariant of the accelerogram is given in terms of the Accelerogram Index $\Sigma (a \Delta t)$ and t, which are the fundamental parameters for this set of data. The invariant of the isoseismal chart is given in terms of the Intensity Index $\Sigma (IS)$ and S, which are the fundamental parameters for this set of data. Furthermore, as stated, the invariants and parameters should be related to *energy*, since this is the essential physical element of the earthquake.

This chapter collected the various theoretical–experimental steps that performed the tasks outlined above. These are represented in the seven curves included in this chapter which are the fundamental data sources for the application of the present methods to the earthquake engineering damage assessment and structural design problems. Before listing these charts and curves, we point out once more (for emphasis), that

1. The acceleratiogram invariant and parameters relate to the timewise variation of local conditions at a point in the earthquake field.
2. The isoseismal invariant and parameters relate to spacewise variations of the total earthquake field.

Both (1) and (2) are reflected in the derived curves as follows:

1. See Eq. 1 and Figure 4.1, which represent the accelerogram invariant. Note the Accelerogram Index $\Sigma(a\,\Delta t)$ and t are the parameters.
2. See Eq. 2 and Figure 4.2, which represent the isoseismal invariant. Note the Intensity Index $\Sigma(IS)$ and S are the parameters.
3. See Eq. 9 and Figure 4.3, which represent an extension of the accelerogram invariant and give the time variation at a point of the surface horizontal energy.
4. See Eq. 14 and Figure 4.4, which represent an extension of the isoseismal invariant and give the spacewise variation, over the entire field, of the surface horizontal energy, W.
5. See Figure 4.5, in which the basic accelerogram parameters $\Sigma(a\,\Delta t)_f$ and t_f are related in a tentative, preliminary fashion to the intensity contours and to the geology. Also shown on Figure 4.5 are general ground energy values for the isoseismal contours. This follows from the hypothesis that, physically, the isoseismal contours must have a quantitative connection with the ground energy, ε_{t_f}. Just what this connection is will be more firmly established as more and more damage data are collected. Essentially, a 'calibration' of the contours is required.
6. See Figure 4.6, in which the basic parameters of the isoseismal invariant, the Intensity Index $\Sigma(IS)_f$ and S_f are shown to be related to the magnitude, M, of the earthquake. Also shown is the effect of geology, based upon the limited available data. As more isoseismal chart–earthquake magnitude data becomes available, this figure also will be revised and refined and corrected. However it appears to be suitable for temporary use as it stands.
7. Figure 4.6 is the fundamental tool for preparation of the MID charts — these being the theoretical isoseismal curves based upon the invariant, Eq. 2. These charts, Figure 4.7, along with Figure 4.5 represent the major source for damage assessment using the theories of this text. Their use will be discussed in more detail in the next chapter.

REFERENCES

1. E. Rosenblueth. (Ed.), *Design of Earthquake Resistant Structures*, John Wiley and Sons, New York, Toronto, 1980.
2. N. Newmark and E. Rosenblueth. *Fundamentals of Earthquake Engineering*, Prentice-Hall, Inc., Englewood Cliffs, N.J. 1971.
3. S. F. Borg. *Extended Analyses of Isoseismal–Magnitude–Intensity Index Correlations in Earthquake Engineering*, Tech. Rept. ME/CE–81–1, Department of Mechanical Engineering (Civil Engineering), Stevens Institute of Technollgy, Hoboken, N.J., May, 1981.
4. S. F. Borg. *A Rational Theory for Predicting Earthquake Damage*, 1982 Annual Convention, American Society of Civil Engineers, New Orleans, Oct. 1982.

5

Approximate Analytical Damage Assessment Procedures

INTRODUCTION

In this chapter the charts, curves, and equations derived from the basic invariant relations are utilized to assess probable approximate damage due to an earthquake. The measure of damage is the 'intensity number' which, by virtue of the Modified Mercalli (MM) Scale (see Chapter 3) will describe greater degrees of damage for different structures as the numbers increase. Once more, it must be emphasized, the results can only be considered as approximate. Quantitatively, some of the values are probably no better than ± 50 per cent or more although others are certainly more accurate. However, in view of the usual accuracy in earthquake engineering analyses, this is perhaps as good as one can expect for engineering design and analysis purposes.

The basic equations, curves, and charts are presented first. Following this several of the different approaches that may be used are described. Basically, one must assume a particular initial set of conditions and from that point various approximate damage-intensity data may be determined, depending upon the needs of the user. In all cases, the invariants and parameters obtained in Chapters 2 and 3 are the fundamental tools used.

THE BASIC EQUATIONS, CHARTS, AND FIGURES

In our analysis of damage we shall utilize the following previously developed equations and charts:

From Chapter 2:

1. The invariant

$$\frac{\Sigma\,(a\,\Delta t)}{\Sigma\,(a\,\Delta t)_f} = e^{0.12[1-(t_f/t)^{1.8}]} \qquad (1)$$

and the plot of Eq. 1, shown qualitatively in Figure 2.3.

2. The point A of maximum acceleration, from Figure 2.3.

$$\left(\frac{t}{t_f}\right)\bigg|_A = 0.25 \tag{2}$$

3. The value of the ordinate on Figure 2.3 corresponding to point A,

$$\frac{\Sigma\,(a\,\Delta t)}{\Sigma\,(a\,\Delta t)_f}\bigg|_A = 0.29 \tag{3}$$

From Chapter 3:

1. The invariant

$$\frac{\Sigma\,(IS)}{\Sigma\,(IS)_f} = e^{2.0[1-(S_f/S)^{1/3}]} \tag{4}$$

and the plot of Eq. 4, shown qualitatively in Figure 3.3.
2. The point A of zero curvature on Figure 3.3.

$$\left(\frac{S}{S_f}\right)\bigg|_A = 0.125 \tag{5}$$

3. The value of the ordinate on Figure 3.3 corresponding to this point A,

$$\frac{\Sigma\,(IS)}{\Sigma\,(IS)_f}\bigg|_A = 0.135 \tag{6}$$

From Chapter 4:

1. The intensity–accelerogram–geology chart, shown qualitatively in Figure 4.5.
2. The magnitude–$\Sigma\,(IS)_f$–S_f correlation, shown qualitatively in Figure 4.6.
3. The magnitude–intensity–distance (MID) chart, typically shown in qualitative form in Figure 4.7.
4. The magnitude–energy equation, Eq. 15 of Chapter 4

$$\mathrm{Log}_{10}\,E \cong 11.4 + 1.5M \tag{7}$$

5. The energy–distance relation,

$$\frac{W}{H} = \frac{\left[\left(\frac{S_f}{S_i}\right)^{1/3} - \left(\frac{S_f}{S}\right)^{1/3}\right]}{\left[\left(\frac{S_f}{S_i}\right)^{1/3} - 1\right]} \tag{8}$$

This equation is shown plotted in Figure 4.4 for sevesal different values of S_i/S_f.

The Determination of ε_{t_f} for the MID Charts

In connection with the determination of ε_{t_f}, the energy per unit area for the MID cusves of Step 3, Figure 4.7, its numerical value depends upon evaluations of the following preliminary quantities:

1. The magnitude–energy relation, Eq. 7.
2. The efficiency of the earthquake, η.
3. The value assumed fos S_f/S_i in Eq. 8.

All three of these require assumptions based upon experience and judgement of the engineer. Following these assumptions, the value of W in Eq. 8 can be determined and from this it is possible to obtain an average value for ε_{t_f} at the various intensity contours. An essential step in the damage assessment and structural design process as developed herein is the determination of the connection between ε_{t_f} and intensity number. It is very possible that all three of the quantities 1, 2, and 3 above have different values for different geologies, magnitudes, and other factors. This will then determine the approximately fixed ε_{t_f}–intensity relation. But a key step is the determination of the actual *numerical* values for ε_{t_f}. At this point, three possible procedures are suggested:

1. Initially, and until a backlog of design data is accumulated, use the equations and recommendations in the previous chapters, including
   (a) The magnitude–energy relation, Eq. 7.
   (b) The efficiency value, $\eta = 5$ per cent–10 per cent (say).
   (c) The S_f/S_i value of Eq. 14, Chapter 4.
   (d) Effective base areas and effective lengths as in Chapter 7.
   (a), (b), and (c) are to be adjusted as required to give reasonably consistent ε_{t_f}–intensity values for different magnitudes and geologies. Various typical structures can be test-designed and the stress–deflection results correlated with the intensity descriptions. This should indicate adjustments for the terms (a) to (d) above.
2. Assume *standard structures* and correlate the energy–stress–intensity relations for these. As an example, a concrete structure of given dimensions will have a computed maximum stress for different combinations of (a) to (d). These results can then be correlated — approximately — with the standard MM intensity scale. In this way, an ε_{t_f}–intensity calibration can be obtained.
3. An experimental procedure might use the data from underground blasts of known energy value at known distances from either a standard structure or simple measuring devices. These results may then be scaled upward to the energies and distances of typical earthquake events.

APPROXIMATE DAMAGE (INTENSITY) ESTIMATES

The damage which a building sustains due to an earthquake depends upon (in addition to other things) magnitude of earthquake, geology of the region, the acceleration record and distance from epicentre. These factors will be included in the damage assessment analysis which will be described in this section. But it must be pointed out that there are other important factors that are extremely difficult (or impossible) to include in an *a priori* determination of overall damage to a particular structure. These include the orientation of the structure. Many structures have strong axes and weak axes and damage depends upon how these are located with reference to the direction of the energy shock. Also the soil–foundation interaction, which depends upon the type of building foundation in addition to soil conditions, affects the overall damage to a building. Finally, the building design (type of construction) affects the damage.

In the later analysis chapter methods will be described that account for, approximately, these different conditions. In this chapter, however, the damage criteria will be based upon the intensity as measured by the Mercalli Scales and is approximate to this degree in addition to the other approximations involved.

Figure 4.5 and, typically, Figure 4.7 are the two charts that are utilized in the damage assessment analysis. These are based upon the invariants (and their extensions) of Chapters 2 and 3, as indicated in earlier sections of this chapter and in the following outlines of the procedures used.

We shall consider some of the different initial conditions assumed by design engineers, code-makers, insurers, and others involved in damage assessment, and how these are then utilized with the invariant relations and charts to determine approximate intensity numbers and, therefore, probable damage.

A. Assume a given magnitude M, geological region, and distance from the epicentre, and efficiency η.
   Procedure: 1. From Figure 4.6, the M–$\Sigma(IS)_f$–S_f correlation, determine $\Sigma(IS)_f$ and S_f.
   2. Determine the proper MID chart, Figure 4.7, for the given M and $\Sigma(IS)_f$ and S_f and η.
   3. From this chart, with the given S, determine the approximate intensity and hence the damage.

B. Assume a_{max} and t_f for a canonical accelerogram and the geology are specified, based upon previous experience and code requirements.
   Procedure: 1. From Eq. 2, determine $t_A = 0.25 t_f$.
   2. Based upon an analysis of several canonical accelerograms, the following parabolic area relation is approximately true,

$$\Sigma(a\,\Delta t)|_A = \frac{a_{max} t_A}{2.8} \tag{7}$$

   3. From Eq. 3 determine

$$\Sigma(a\,\Delta t)_f = \frac{\Sigma(a\,\Delta t)}{0.29}\bigg|_A$$

$$= \frac{(a_{max})(0.25)(t_f)}{(0.29)(2.8)} \tag{8}$$

4. Having t_f and $\Sigma(a\,\Delta t)_f$ and the geology, from Figure 4.5 determine the approximate intensity, and hence the damage.

C. Assume that a composite accelerogram, consisting of several superposed canonical accelerograms each one similar to (B) above is given. Also the geology.

Procedure: If the accelerogram which is specified consists of a series (summation) of canonical accelerograms as in Figure 2.5 of Chapter 2, then it is recommended that each canonical part be considered separately insofar as the $\Sigma(a\,\Delta t)_f$–t_f–Intensity Damage Chart is concerned. That is, use Figure 4.5 for each separate canonical part and determine the worst case approximate intensity and hence the damage.

This is justified by the physical fact that many of the canonical segments appears to be practically independent of each other with, at most, a slight overlap at the head and tail (low acceleration) ends of each. Thus insofar as local effect (intensity) is concerned, each part is essentially independent. If the superposed accelerogram consists of segments which overlap more completely, a modification of the procedure will be indicated.

Another form of design information that may be related to a particular earthquake and region is the following.

D. Assume a structure is designed so that it is capable of absorbing a given amount of energy about major and minor axes.

Procedure: 1. From the MID charts (essentially a handbook of M, η, ε_t, $\Sigma(IS)_f$, R and S_f values), it is possible to determine the total etnergy, $\varepsilon_{t_f}(Ae)$ that a structure must absorb. Ae is the 'effective area' about which more will be written in later chapters. It is thus possible to determine various combinations of M, η, $\Sigma(IS)_f$, S_f, geology, and S corresponding to required total energy correlated with the possible orientations of the building. If the orientation is unknown then prudence would require that the weak axis direction be required to absorb the entire energy supplied by the earthquake.

There are other combinations of the variables and the charts that can be used as the basis for a given engineering or damage or code analysis. This depends upon the needs of the person making the study. The four examples given above are typical and indicate the procedures that may be used.

CONCLUSION

The invariants of the accelerogram and isoseismal chart, and the different extended relations, curves, and figures that follow from these, which were developed in previous chapters are used as the basic tools for determining — analytically — approximate damage to structures due to an earthquake.

The damage criterion is the Mercalli Scale intensity (MM) number and most of the more important variables involved in damage phenomena are accounted for. The procedures are approximate and depend upon a compilation of damage data from previous earthquakes, i.e. basically a calibration of the curves and charts. As more and more data is collected and recorded, the basic design tools should become more accurate and can be used with greater confidence. As pointed out in Table A.2 of the Appendix, the determination of this data (which has a specialized form differing markedly from more familiar experimental structural design data) is entirely consistent with procedures used in other fields of applied mechanics, as required because of the fundamental uniqueness of earthquake engineering within the overall field of applied mechanics.

Finally, other combinations of input and output data than those given as examples in this chapter are possible and may be obtained. These depend upon the needs of the engineers, insurers, code-makers, and others who may find these data of interest.

6

Special Topics in Earthquake Structural Engineering

INTRODUCTION

Before proceeding to the discussion of the structural analysis and its application, it is desirable to introduce four special topics that are related to the analysis and design and have important connections with these subjects. The four topics are:

1. Equivalent length and equivalent base area.
2. Length-of-time effect.
3. Damping.
4. Model–prototype relations.

The connections of the four with structural analysis will be emphasized and in common with all other sections of the book, arguments will be based primarily upon engineering and physical considerations.

Several new concepts will be introduced. These also will have engineering–physical bases and will be significant elements in the overall theory to be described in the remaining chapter.

NOMENCLATURE FOR THIS CHAPTER

Following is a list of the terms used in this chapter. They will also be defined when they are introduced.

Ae = equivalent area
g = acceleration of gravity
l = equivalent length
l' = actual length
t = time
w_o = inertia loading on structure
x = lateral deflection of structure
y = distance along axis of structure
A = average uni-material area of building
E = modulus of elasticity

F = subscript 'framing'
G = modulus of rigidity (shear modulus)
I = moment of inertia
M = bending moment
P = load acting on base of building
S = subscript 'shell'
U = strain energy
V = velocity of shear wave
ε_{t_f} = total horizontal ground energy per unit area
σ = stress
ρ = density
ω = frequency

EQUIVALENT LENGTH AND EQUIVALENT BASE AREA

For illustrative purposes in this chapter we will consider a building 100 ft × 100 ft in cross-sectional area having an 'actual' length, l', as shown in Figure 6.1. Note that l' is the true physical length of the building from roof to bottom of footings;

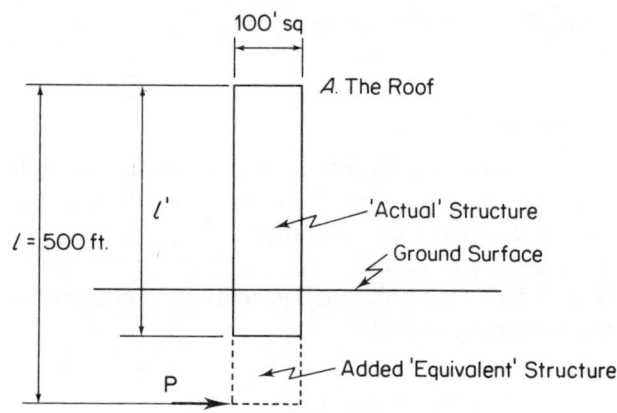

Figure 6.1

$l = 500$ ft is the 'equivalent' length (height) of the building, assumed known; l is the length actually used in the computations and is the length that accounts for the soil–foundation interaction as well as for the translation and rotation of the actual base of the building. This is an important (and frequently neglected) aspect of the structural response of a building.[1]

In this text, by using l instead of l', the translation and rotation of the actual base are introduced as normal elements of the design procedure. The determination of the distance l is based upon the judgement and experience of the designing engineer as well as upon an empirical relation for the period of the building as discussed in the next chapter. Note also, the earthquake loading (shown in Figure 6.1 as P) is applied at the base of the equivalent structure, l. The portion of the structure, l–l' accounts for the soil–foundation interaction corresponding to the actual structure beneath the ground surface. This also is clearly an item strongly dependent upon engineering judgement.

A second important consideration in the response analysis is the 'equivalent base area', Ae, which when multiplied by ε_t (see Eq. 6 of Chapter 4) determines the amount of energy from the earthquake which the structure must absorb. Ae may be larger or smaller than the actual base area (100 ft × 100 ft in this case), depending upon the geology and soil conditions and perhaps other factors as well.

One of these additional factors which almost certainly has some important bearing on the value to be assigned to Ae is the total superposed dead load and live load of the structure since this very probably is related to the energy which the structure must absorb. That this is so may be justified by the following physical argument:

A superposed load is equivalent (insofar as the underground is concerned) to an additional ground overburden and therefore the actual ground surface is essentially at a position lower than it actually is. The heavier that the superposed load is, the greater is the effective lowering of the ground. This lower position (footing elevation) means that a greater energy must be absorbed by the building, a determination which would be included in the assigned value of Ae for the given structure by the engineer or code.

As more and more designs are considered and a backlog of experience and data are accumulated, this term (and also l), will be determined with greater confidence and accuracy. Typical test projects (such as underground blasts) can be designed for known given conditions and these will give some hints concerning reasonable values of Ae as well as l. See related comments in the Conclusion of Chapter 5.

As a first approximation (and until more data is available) it is suggested that the actual building base area be used as Ae. Whether or not this is reasonable can be determined by comparing the design results with the damage assessment prediction obtained as outlined in the previous chapter.

This point must be emphasized — values of l and Ae, initially, will be based upon the experience and judgement of the designer as are the evaluations of various terms that occur in earthquake codes throughout the world. Perhaps the specifications will suggest a suitable range of values for these. As more data is stored and accumulated, it will be possible to determine more accurate and reasonable values for these two important design parameters.

LENGTH-OF-TIME EFFECT

Referring to Figure 6.1, note that l is the equivalent length, i.e. the length used in the analysis. The earthquake load is shown as P acting on the base of the l structure. Also shown are the dimensions considered for illustrative purposes.

We assume the velocity of sound (i.e. the velocity of, in this case, shear stress and displacement effects), is 10,000 fps in steel and 5000 fps in brick, mortar, and concrete. These could vary depending upon several factors, but in any case there will be, roughly, a factor of two between the steel and brick–concrete values. We assume further that the load due to the earthquake is initially, the force P.

The effect of P will travel up the building and will be felt at the roof, A, in about 1/20 s, insofar as the steel frame is concerned.

1. We shall assume, in this discussion, that the structure of equivalent length, l, vibrates as a free–free beam. We shall assume, therefore, that the entire building of length l, vibrates as a free–free beam for time periods, t_3, given by, roughly

$$\frac{1}{20}\text{s} < t_3 \qquad (1)$$

This time period, therefore, is called the *period of overall effect*.

2. If the building cross-section is assumed to be 100 ft × 100 ft then the force, P, applied as shown, will require approximately 1/100 s to traverse the building width, i.e. to be felt at the far side of the building. That is, for approximately

$$0 < t_1 < \frac{1}{100}\text{s} \qquad (2)$$

the stress and deformation characteristics of the earthquake loading are predominantly localized (in floor slabs, beams, girders, wall panels in the neighbourhood of the load application point) and can only be accounted for by considering the separate effects on these structural elements and, very likely, the three dimensional equations of dynamic elasticity. The problem in this last case is an extremely complicated one and can be solved, if at all, only for very simplified, idealized structures.

The time period

$$0 < t_1 \leqslant \frac{1}{100}\text{s} \qquad (3)$$

is, for the reasons stated, called the *period of localized effect*.

3. The intermediate time period

$$\frac{1}{100}\text{s} < t_2 < \frac{1}{20}\text{s} \qquad (4)$$

is called the *period of transition* in this book. During this time interval, the entire building is still not aware of the earthquake. Furthermore, for those buildings with a steel skeleton framing and a brick outside wall, it is possible that severe damage and wall cracking can occur during this time period because of the difference in the velocity of sound in the two materials as explained in the following:

Suppose $t = 1/50$ s. During this time, the wave will have travelled roughly, 200 ft up the *steel framing portion of the building*, see Figure 6.2, with the

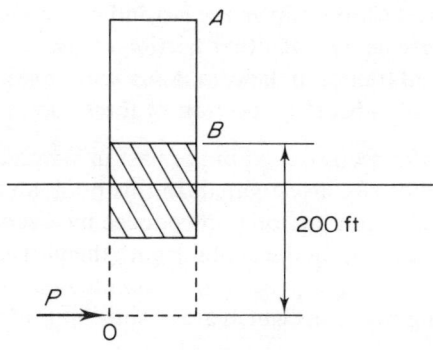

Figure 6.2

corresponding shear deformation. On the other hand, the wave through the brick will only have travelled about 100 ft, so that in the upper 100 ft there is deformation of the steel without any deformation of the brick. This could lead to serious cracking and damage problems unless the two materials can deform independently of each other.

In effect, the portion of the building shown shaded in Figure 6.2 is aware of the earthquake (i.e. is deforming, is being stressed). The portion AB of the building is unaware of any earthquake loading at point 0.

The three time periods described above appear to be the significant ones in earthquake response analyses. They will be utilized in this book's consideration of the effect of earthquakes on structures.

DAMPING OF VIBRATIONAL BUILDING OSCILLATIONS DUE TO AN EARTHQUAKE

When a building or other structure is subjected to a sudden impulse-type earthquake loading, it is set in vibratory motion. A convenient way of examining the response of the structure is to introduce energy considerations, that is, to assume that the vibrations are due to an energy input of known calculated value, which in turn induces kinetic and potential (strain) energy in this structure. This is the method utilized in this text (see Chapter 7) and the same considerations may be utilized in investigating the damping behaviour of the structure.

In this book therefore, we conceive of an earthquake as a mechanism for introducing energy into the system (the structure, including soil–substructure interaction). If damping were absent, the resulting vibrations would exist for all time. Because of damping, energy is withdrawn from the system, which in turn reduces the amplitude of vibration until the final stationary state is reached.

As a first step it is necessary that the damping mechanism be described as completely as possible. There seem to be three possible major modes of damping involved in the dissipation of energy from a vibrating structure, as follows:

1. Energy lost, due to internal friction, in the vibrating building shell, frame, partitions, floors, etc. subjected to alternating stress cycles.

2. Energy lost, due to friction and slip, in the lap and butt riveted–bolted joints of the vibrating structure as well as other frictions between contiguous parts.
3. Energy lost as internal friction in (and in doing work against) subsurface soil by the foundation and subsurface portion of the structure.

Items (1)[2] and (2)[3] have been studied in the past in simplified forms. Item (3), however, has been only briefly investigated. In this book an approximation for energy lost in soil–structure interaction is introduced by assuming an 'equivalent length', l, of structure greater than the actual length (height) as shown in Figures 6.1 and 6.2. The equivalent length procedure as pointed out before enables us to accomplish the following two tasks which are of some importance in earthquake structural analyses:

1. By assuming an equivalent length structure we introduce translation and rotation of the actual foundation. These motions may be important or even critical in certain cases and should not be overlooked.
2. By considering an equivalent length we introduce an *added* energy loss in the vibration cycle. This loss (damping) is assumed to represent the effect of the soil–substructure effect on damping.

The difficulty with item (3) analyses is the determination of l, the 'equivalent' length. This obviously must depend — among other things — on the type of soil, the type and size of foundation, and the intensity of the shock. One suggested approach will be to analyse structures utilizing variable equivalent lengths, preparing tables and charts, and attempting to correlate these with actual behaviour. If such a correlation can be obtained and this is sufficiently accurate for engineering design purposes, then a relatively simple technique will be available for including soil–substructure interaction effects in the design of structures subjected to earthquake loads.

Until this information is available, the author suggests that the equivalent length be determined by using the empirical period of vibration relation as outlined in Chapter 7 modified, if deemed necessary, by engineering judgement and experience.

MODEL SCALING REQUIREMENTS

Introduction

The earthquake phenomenon is an extremely complicated one involving, as it does, dynamic as well as elastic (possibly plastic) effects. In connection with the model testing of structures subjected to earthquake loadings, the questions to be answered initially are what the scaling requirements of model to prototype must be in order that the results be meaningful.

To answer these questions, it is necessary that the complex behaviour of the structure be categorized or separated into phenomena that can be analysed. This section reports on a study along these lines.

It must be stated at the outset, that in this analysis we are concerned with the overall scaling requirements for the structure. We do not consider details such as connections or isolated members. Many studies have been reported along these lines and the scaling requirements for such cases have been established. Our concern is with a set of requirements that are less well known, generally ignored and therefore lead to questions concerning the validity of the results of model tests.

We begin by discussing the basic technique to be utilized in the analysis. Following this, the separate effects which are to be considered are introduced and, in general, given in their dimensional form. This, in turn, leads directly to the required scaling parameters. We include a brief discussion of the scaling factors involved in vibrational damping. Finally, the question — 'Is it possible to simultaneously satisfy all the scaling parameters?' is discussed briefly.

Procedure

One method which is very fruitful in model and scaling analyses utilizes the concept of 'ratio of effects'. That is, we may say that a model will behave the same as a prototype if the 'ratio of two effects' is the same. For example, in classical fluid dynamics, this approach leads to the following:

1. If, for a prototype, inertia and viscous effects (forces) are the important elements in the phenomenon, then, for model results to be meaningful, we require that the ratio of inertia effects to viscous effects be equal for both model and prototype. That is, we require the Reynolds number be the same.
2. Or, if for model and prototype, we require that the ratio of compressibility effects to inertia effects be equal, then we require that the Mach number be the same.
3. Or, if for model and prototype, we require that the ratio of surface-wave effects to inertia effects be equal, then we require that the Froude numbers be the same.

And so on for other 'ratio of effect' combinations.

We shall use this concept in our analysis of the earthquake model–prototype relationship. The significant 'effects' in our problem shall be assumed to be the following geometric, elastic, and dynamic terms.

1. The shear-bending wave transmission characteristics of the building — both steel framing and brick or concrete or other outer covering.
2. The lateral wave (vibratory wave) characteristics of the building.
3. The 'effective length–actual length' effect.
4. The time-loading history of the earthquake load.
5. The strain energy absorbed by the structure.
6. The total energy supplied to the structure by the earthquake.
7. The maximum stress induced in the structure.
8. The maximum allowable stress (modulus of rupture or yield stress) of the structure.

Our criterion is the following: in order that the model be scaled properly with respect to the prototype insofar as overall, major earthquake effects are concerned, it will be necessary that the 'ratio of effects' of the foregoing quantities be the same for model and prototype.

We take the point of view that we wish to scale our model in such a way that its complex behaviour under an earthquake loading will be similar to that of the prototype. In this way, by means of model studies, we will be able to predict how a structure will behave in an earthquake and also be able to test the validity and applicability of any theory for the effects of earthquakes on structures.

The first four of the above effects are important, because they mean that the various *phase relations* of the lateral deflection wave, bending wave, earthquake loading time-history, and so on, for the model should be as nearly the same as possible as those of the prototype. To illustrate, in a purely physical and phenomenological way, the significnace of the various relations, let us examine the behaviour of a typical building structure subject to a short-pulse earthquake loading, Figure 6.3.

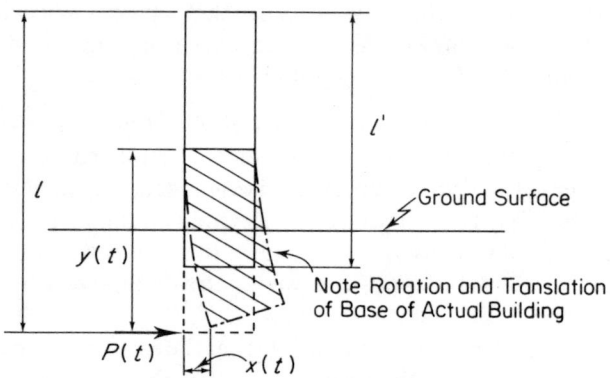

Figure 6.3

First, based upon soil–substructure interaction and other pertinent data, an effective length, l, is assumed for the building. At the instant the earthquake load $P(t)$ is applied, a shear wave begins to travel up the building (shown by the distance $y(t)$) and a lateral shear deflection occurs as indicated by the distance $x(t)$. If the model is to behave in a manner similar to the prototype, then it is necessary that:

1. The ratio l'/l must be the same for model and prototype.
2. The ratios of the different distances $y(t)$ for steel framing and for brick shell (say) must be the same for model and prototype.
3. The ratio of the shear deformation, $x(t)$, to distance of shear wave travel, $y(t)$, must be the same for both model and prototype for a given ratio $y(t)/l$.
4. The ratio of the critical frequency of the applied load to the frequency of lateral vibration must be the same for model and prototype. 'Critical' frequency will be discussed in the following paragraphs.

In addition to the above ratios, the following also must hold for model and prototype:

5. The ratio of maximum strain energy in the structure to total energy applied to the structure by the earthquake must be the same in the model and in the prototype.
6. The ratio of the maximum stress in the structure to the modulus of rupture (or yield stress) must be the same for model and prototype.
7. The damping ratio must be the same for model and prototype.

The above seven geometric, elastic, and dynamic scaling requirements should be satisfied if the model test results are to represent a true picture of the prototype structure. They show clearly that one common method of testing, i.e. using a 'shaking machine' on the roof or some other floor of a model (or, indeed, prototype) cannot possibly give results corresponding to what actually occurs during an earthquake, since the earthquake load — of necessity — is applied by the ground and the base of the building almost certainly rotates and translates during the time of loading. Unless these last noted motions are introduced — in both test or analytical solution — an important (possibly critical) element of the phenomenon has been omitted.

We now discuss the seven scaling requirements in more detail. The number headings conform to those used above.

1. This requires that

$$\left(\frac{l'}{l}\right)_{model} = \left(\frac{l'}{l}\right)_{prototype} \tag{1}$$

2. The velocity of a shear wave is given by

$$V = k\left(\frac{Gg}{\rho}\right)^{1/2} \tag{2}$$

in which G = shear modulus of elasticity, g = acceleration of gravity, and ρ = density of material.

The scaling requirements are therefore

$$\left\{\frac{\left(\frac{Gg}{\rho}\right)^{1/2}_{shell}}{\left(\frac{Gg}{\rho}\right)^{1/2}_{framing}}\right\}_{model} = \left\{\frac{\left(\frac{GG}{\rho}\right)^{1/2}_{shell}}{\left(\frac{Gg}{\rho}\right)^{1/2}_{framing}}\right\}_{prototype} \tag{3}$$

or

$$\left(\frac{G_S \rho_F}{G_F \rho_S}\right)_{model} = \left(\frac{G_S \rho_F}{G_F \rho_S}\right)_{prototype} \tag{4}$$

3. It is assumed in this analysis that the deformation during the period of transition is a pure shear translation only. That is, bending effects have not been introduced until the lateral vibration of the structure begins. (This point will be discussed in more detail in the next chapter.) Thus, in Figure 6.3, the distance $x(t)$ is given by

$$K \frac{Py}{AG} \tag{5}$$

in which

K = a numerical factor to account for suddenness of load, form factor, etc.
P = constant pulse load.
y = distance of shear wave travel.
A, G = average uni-material area and modulus of rigidity.

Then, for a given $y(t)/l$, the scaling requirement is that

$$\frac{Py}{AG/y} = \left(\frac{P}{AG}\right)_{model} = \left(\frac{P}{AG}\right)_{prototype} \tag{6}$$

Subject to

$$\left(\frac{y}{l}\right)_{model} = \left(\frac{y}{l}\right)_{prototype} \tag{7a}$$

or

$$\left(\frac{Gt^2}{\rho l^2}\right)_{model} = \left(\frac{Gt^2}{\rho l^2}\right)_{prototype} \tag{7b}$$

4. The frequency equation for the free–free beam modes of vibration (which is the assumed condition for the equivalent building in this formulation) is given by[4]

$$\omega = K\left(\frac{EIg}{\rho Al^4}\right)^{1/2} \text{ rad s}^{-1} \tag{8}$$

in which

ω = frequency in rad s^{-1}
E = tension–compression modulus of elasticity of uni-material building
I = moment of inertia of equivalent uni-material building cross-section
g = acceleration of gravity
ρ = density of equivalent uni-material building

A = cross-section area of equivalent uni-material building
l = equivalent length of building

In Chapter 2, the study of earthquake accelerograms showed that, for many earthquakes, the accelerograms can be represented in an invariant canonical form when analysed between envelopes and for suitable choice of invariant-form variables. Furthermore, for those accelerograms not directly representable by the canonical invariant, it is likely that a superposition of these may be utilized to represent the more complicated accelerograms.

It will be assumed in this book that a possible loading on the building due to the earthquake consists of a series of short pulses, acting in alternate directions, bounded by the shape of the envelopes of the accelerograms and having *a most critical frequency insofar as the response of the structure is concerned*. The pulses, the envelopes, and a typical simplified accelerogram is shown in Figure 6.4.

Note the pulse loads have a particular frequency (which may or may not be assumed constant). There is one frequency spectrum which is probably critical for a given building, in that it will cause the most damage to the building. Let us call this frequency ω_{crit}. Therefore the scaling requirement for load application requires that the ratio of critical load frequency to transverse vibration frequency be the same for model and prototype, i.e.:

$$\left(\frac{EIg}{\rho A l^4 \omega_{\text{crit}}^2}\right)_{\text{model}} = \left(\frac{EIg}{\rho A l^4 \omega_{\text{crit}}^2}\right)_{\text{prototype}} \tag{9}$$

5. The maximum bending strain energy in the vibrating structure, see Chapter 7,

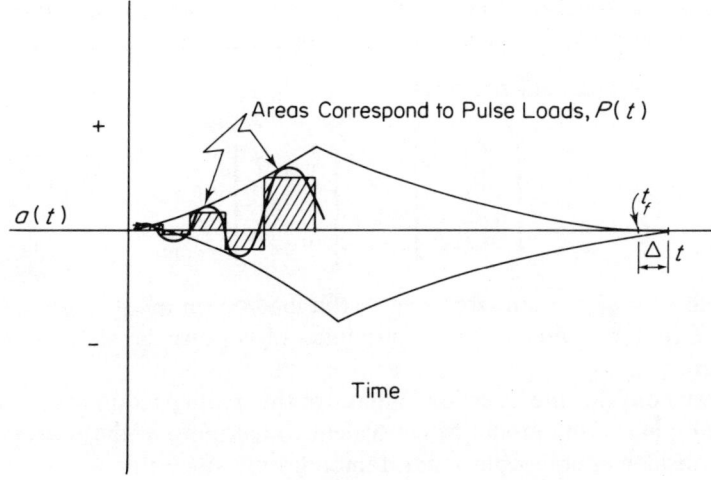

Figure 6.4

will be given by an expression of the form

$$\int \frac{M^2 \, dx}{2EI} \tag{10}$$

in which M = bending moment at any point on the vibrating structure due to the potential (inertia) load, E, I are as defined in Eq. 8, and the integration is taken over the entire free–free vibrating structure.

The total energy, due to the earthquake, that must be absorbed by the structure is given by $\varepsilon_{t_f} Ae$, in which

ε_{t_f} = total horizontal ground energy per unit effective area supplied by the earthquake.

Ae = effective base area of the structure.

The model–prototype requirement is, therefore,

$$\left[\frac{\int \frac{M^2 \, dx}{EI}}{\varepsilon_{t_f} Ae} \right]_{model} = \left[\frac{\int \frac{M^2 \, dx}{EI}}{\varepsilon_{t_f} Ae} \right]_{prototype} \tag{11}$$

As noted above, this relation assumes *bending* energy alone is the significant quantity. Depending, for example, upon the geometry of the structure it is possible that shear energy alone or a combination of shear and bending energy are important and the relation Eq. 11 would therefore be modified accordingly.

A complication that arises in connection with this requirement has to do with the major–minor axes directions of the structure versus the direction of the ground motion (which, in this text, is assumed to be in a radial direction from the epicentre to the structure).

The most reasonable procedure might be to consider the direction of maximum effect and to look for an approximate equality for Eq. 11.

6. The stress requirement becomes

$$\left[\frac{\frac{w_0 l^2 d}{I}}{(\sigma_{r \text{ or } y})} \right]_{model} = \left[\frac{\frac{w_0 l^2 d}{I}}{(\sigma_{r \text{ or } y})} \right]_{prototype} \tag{12}$$

in which w_0 = maximum vibratory inertia loading on the structure, d = depth of the structure, and $\sigma_{r \text{ or } y}$ = the modulus of rupture or yield stress of the structure.

7. Next, we consider the question 'What are the scaling requirements in order that damping of the model be equivalent to damping of the prototype? We shall consider lateral vibrational damping only since this is very likely the major source of damping and — in addition — damping due to soil–

substructure interaction is accounted for approximately, in this theory, as added vibrational damping due to the additional length in the equivalent length formulation.

In discussing vibrational damping,[2] it is shown that the energy damped per cycle is given by

$$\Delta U = 2\pi \, \delta U \tag{13}$$

in which ΔU = energy lost per cycle, U = maximum strain energy in the vibrating structure, i.e. when the structure is at its maximum deflection, and δ = damping factor.

Furthermore, it is shown[2] that although δ is dependent upon a number of factors, its main dependence, for our purposes, for any given material is upon ω, the frequency of vibration. Also — and this may be important for model-scaling purposes — it is shown that ω is frequently a double-valued function, i.e. any given value of δ will occur at two different values of ω.

There are a number of approximate relations giving the period of vibration for actual building structures.[5] Hence, if we define equivalent damping to mean

$$\left(\frac{\Delta U}{U}\right)_{model} = \left(\frac{\Delta U}{U}\right)_{prototype} \tag{14}$$

then we require (if model and prototype are constructed of the same materials) that

$$\omega_{model} = \omega_{prototype} \tag{15}$$

in which we have the leeway presented by the double-valued nature of the damping-function.

The final question — 'Is it possible to satisfy all of the above requirements simultaneously?' The answer to this question is dependent upon such a large mix of geometric, vibratory and elastic properties, and physical parameters, that it is unlikely all can be satisfied. Some contradictions almost certainly will arise. The best one can do is to assume some of the requirements are more important than others and attempt to satisfy these.

CONCLUSION

Four special topics dealing with earthquake engineering analyses and experimental work are considered in this chapter.

1. The concepts of equivalent length and equivalent base area are of major importance in the structural procedure developed in the following chapter.
2. The length-of-time effect may help to explain one of the primary causes of

some building damage — a differential deformation of adjoining parts of the structure.
3. The damping of a structure is a basic element in its vibratory behaviour, and one of the more difficult factors to quantify. Some approximate procedures are suggested to account for the responses of structures to earthquakes.
4. Experimental model work in earthquake engineering is an important area of activity. The difficulty with the use of models is that one must be certain that the model response is, in fact, a reasonable indication of the full-size structure response. One means for assuring this is by utilizing the 'ratio of effect' concept and this is carried through to determine various geomtric, elastic, vibratory, and stress requirements.

REFERENCES

1. Nathan M. Newmark and Emelio Rosenblueth. *Fundamentals of Earthquake Engineering*, Prentice-Hall, Inc., Englewood Cliffs, N. J., 1971, p. 94.
2. C. Zener. Elasticity and Anelasticity of Metals, University of Chicago Press, Chicago, Ill., 1948.
3. E. L. Gayhart. 'An investigation of the behavior and of the ultimate strength of riveted joints under load', *Trans. SNAME*, **34**, 1926.
4. N. W. McLachlan. Theory of Vibrations, Dover Publications, Inc., New York, NY, 1951.
5. Robert L. Wiegel. *Earthquake Engineering*, Prentice-Hall, Inc., Englewood Cliffs, NJ, 1970, p. 88.

7

The Structural Analysis Procedures: Symmetry, Anti-symmetry, Energy

INTRODUCTION

As the sub-title of the book indicates, two main aspects of earthquake engineering were considered, these being

1. Damage assessment, and
2. Structural analysis and design.

Furthermore, the entire theory was to be formulated and integrated in a rational fashion on the basis of the two major sets of experimental field data, the accelerogram and the isoseismal chart. The glue or unifying element was to be energy.

In the previous chapters the theory was developed ih terms of parameters and invariants specially related to earthquake engineering and the application to damage assessment was described (Chapter 5). In this chapter the theories and methods of the earlier portions of the book are utilized in obtaining an approximate structural procedure for determining the effects of earthquakes. Here also, energy (in the form of strain energy and vibrational energy) is the basic quantity which determines the adequacy of the structure.

Various assumptions must be made concerning the energy supplied by the earthquake and the energy absorbed by the structure. In a real sense, the theory described in this chapter is in the same position as was the very early strength of materials (or structural analysis) theory. For example, in structural analysis, if we consider beam and column design, we may assume that the flexure formula, shear equation, and Euler column relation are known. However, before a beam or column can be designed (in, say, steel or concrete), it is necessary that *allowable* bending and shear stresses and the modulus of elasticity be known. These can only be determined by actual testing of the appropriate materials. Failing in this, the values must be assumed.

In the same way, in our earthquake analyses, the approximate basic relations were obtained in earlier chapters, but actual *allowables* are needed (in terms of energies that can be absorbed by structures). Also, actual calibrating timewise and spacewise energy distributions are needed for canonical accelerograms, for isoseismal maps, and for different geologies. Without knowing these quantities, they must be assumed, initially.

Evaluations must be made (and recorded as a data source) for standard earthquakes and for standard structures for the different geologies. When this has been accomplished, the design process and the damage assessment analyses will become, in theory, no more complicated than beam and column analysis for simple structures. A more detailed discussion of various points related to the above is contained in the Appendix.

There are a number of key assumptions made in the structural procedure. They will be listed here and discussed more fully, as needed, when introduced in later parts of this chapter.

1. An effective length of structure must be assumed. This is l.
2. An effective area of the base of the structure must be assumed which will determine the amount of energy that the structure must absorb. This is Ae.
3. The structure of length l will be assumed as a free–free beam insofar as bending response to the earthquake is concerned. Short, squat structures ($l/d < 1$ to 2) will be treated as shear responders.
4. For our present purposes, a uniform structure will be considered. In actual cases, the true shape of the structure will offer no complication. Computer programs are available for determining the required design data, such as node locations and principal periods.
5. Loadings and responses will be assumed as separable into symmetric and antisymmetric components and superposable by simple addition.
6. For structures built of different materials, such as a steel frame and brick facing or similar, it will be assumed that a division of energy between the separate parts is known or can be assumed. A uni-material analysis may be appropriate.
7. The earthquake magnitude–energy–efficiency relationship for different geologies will be assumed as known.
8. The temporal and spacewise variations of surface horizontal energy will be assumed as given in the earlier chapters pending collection of actual calibrating data on these items.

This chapter presents, in some detail, the procedure to be used in solving the complete approximate time-history response for a structure (in the elastic range) subjected to an earthquake loading. In the explanation of the method, special attention is given to the physical interpretation of the various steps involved in the proposed solution. The physical explanation and significance of the proposed division between shear-wave effects and bending vibrational effects (which is the foundation of the proposed method) is covered in detail.

Methods of elastic analysis for the dynamic earthquake problem are given in the literature in very general terms, but actual numerical computations are quite rare. It should be noted that what is being discussed in this chapter is the solution for the shear, moment, and deflection — not the frequency solution. Many computations have been given for the frequency analysis — and this is an important physical quantity since resonant vibrations due to rotating machinery and parts can cause very undesirable effects unless properly designed for — and this proper design requires a knowledge of the frequencies of the various modes of vibration. However, the present discussion is concerned only with the analysis of the effects of earthquake loadings. In particular, earthquakes can lead to a considerable increase in the critical bending moment in structures. An approximate method for calculating these added moments (and shears and resulting deflections) is presented in this section.

It must also be emphasized that the method described in this chapter is, admittedly, an approximate one. However, all methods which have been presented as solutions to this problem are approximate to some degree. Also, as the discussion in the Appendix indicates, existing methods of analysis are open to serious criticism with reference to the accuracy of the designs to which they lead.

The present method is a fairly simple one to apply, based on physical reasoning, and one which can be very simply visualized from the engineering point of view.

PROCEDURE

For purposes of simplicity only, the discussion will be referred to a constant section structure, Figure 7.1(a), and the load, initially, will be an assumed pulse-type load as shown in Figures 7.1(a) and (b). However, the method is a general one and can be applied also to variable section structures and to variable loadings. In the case of the variable section, a graphical method of analysis can be utilized. Any given alternating load can be approximated by a summation of alternating pulse-type loads.

The response of the structure-shear, moment, and deflection, due to elastic–dynamic effects caused by the loading, will be obtained starting with $t=0$. Note the phrase 'elastic–dynamic effects': this means that we are not interested in the rigid-body motion of the structure. This could be obtained using Newton's equations and the combined effect is a simple superposition of rigid-body plus

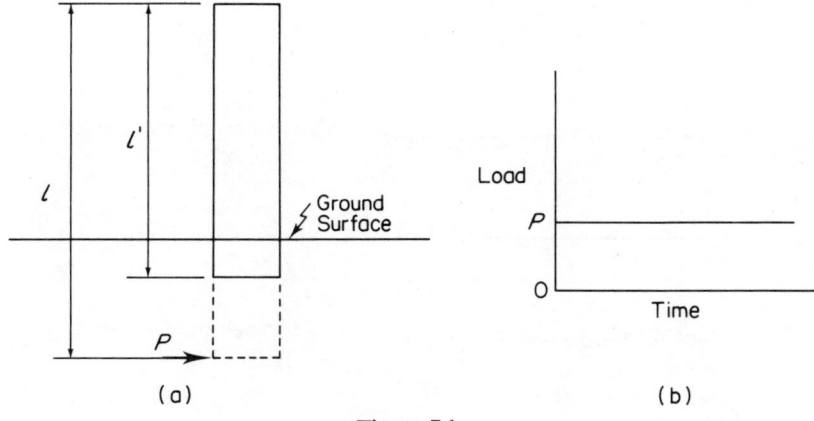

Figure 7.1

elastic-body motions. In this discussion, however, we shall be concerned with the elastic-body effects only.

In Chapter 6 three separate time regions were distinguished. We shall be concerned with the responses only in two of these, namely: (a) the period of transition, and (b) the period of overall effect. The solution in the other region, the so-called period of localized effect, requires a detailed knowledge of the local structure in the neighbourhood of the point of application of the load and is of no interest here, although this could be critical.

The proposed method of solution is based primarily upon energy considerations in so far as (b), the period of overall effect is concerned. However, the response during the time period (a), the period of transition, is based upon the wave-transmission characteristics of the structure, and to some extent has its justification in the actual behaviour of a structure when subjected to earthquake loading. Before considering this point in greater detail, the question of symmetry, anti-symmetry, and unsymmetry will be discussed, particularly with reference to loading effects on the structure.

SYMMETRY, ANTI-SYMMETRY, AND UNSYMMETRY

For a uniform free–free beam section, the classical solution for the normal modes of vibration indicates,[1] alternating symmetrical and anti-symmetrical (or skew-symmetrical) shapes (see Figure 7.2). In our discussion, we shall be concerned only with the four modes shown. If additional modes are needed, they can be included. Or, if a more approximate (quicker) solution is desired, only the first two modes need be considered and in view of the inevitable approximations, we shall assume the two mode solution as acceptable for most cases.

We shall assume, therefore, that in general S2 and S4 are the developed symmetrical modes of vibration, AS3 and AS5 are the developed anti-symmetrical modes of vibration, although for simplicity, S2 and AS3 only will be used in the detailed explanation of the method.

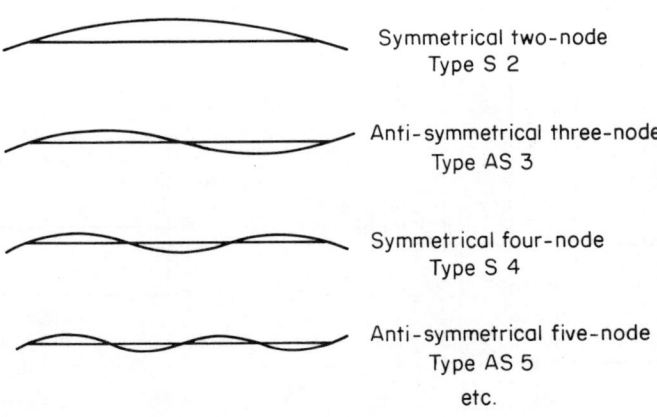

Figure 7.2

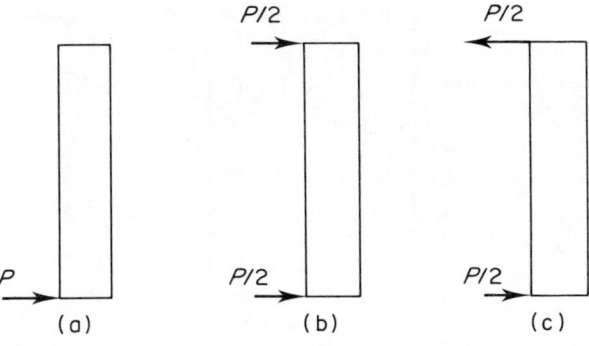

Figure 7.3

With reference to loadings, any arbitrary unsymmetrical load on a beam structure can be replaced by an equivalent pair of loadings, one of which is symmetrical, the second of which is anti-symmetrical. Thus, in Figure 7.3(a), the arbitrary load is an unsymmetrical one. Figure 7.3(b) is a symmetrical loading, Figure 7.3(c) is an anti-symmetrical loading and, obviously, the sum of the cases 3(b) and 3(c) is equivalent to the case 3(a).

We shall assume that the loading and vibratory response can be analysed in terms of symmetry, anti-symmetry, and unsymmetry. This is fundamental in the proposed method of solution. Furthermore, it will be assumed that the overall effect can be approximated as a simple superposition of the separate symmetrical and anti-symmetrical effects.

PERIOD OF TRANSITION

One eyewitness description of an earthquake effect on a tall building is that the initial awareness — immediately after the earthquake is felt — is of a 'shudder' going up the building. Shortly after that, the building begins to 'sway', i.e. vibrate. The proposed method of solution is, physically, one which gives a response similar to that described. The equivalent of the 'shudder' is introduced by means of a shear wave travelling up the length of the building. This shear wave introduces required balancing inertia forces and introduces only shear deflections (although bending moments also are induced, but bending deflections have not had time to develop). The velocity of this shear wave is given by the known formula,

$$V = k\left(\frac{Gg}{\rho}\right)^{1/2} \tag{1}$$

and the equilibrium conditions and loadings shown in Figure 7.4 are consistent with this velocity. Thus, Figure 7.4a is at a time $t = t_2$, where t_2 is within the period of transition defined earlier. Figure 7.4b shows conditions at the instant that the period of transition ends, and Figures 7.4c and 7.4d show Figure 7.4b in terms of

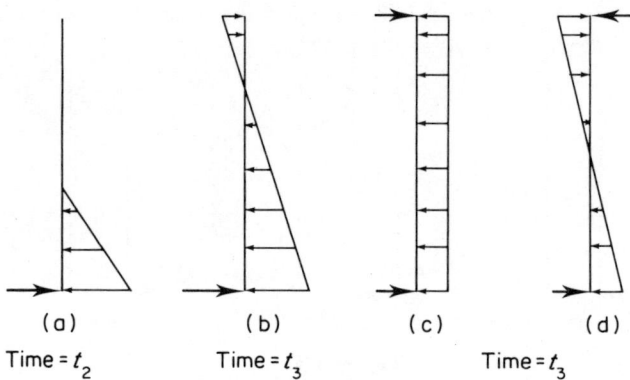

Figure 7.4

the symmetrical and anti-symmetrical loadings. Figure 7.4a is assumed to be the period during which the 'shudder' is travelling up the building. Vibrations corresponding to Figure 7.3 have not as yet begun. The shear and moment at any point during this period can, however, be obtained. The balancing inertia loads are assumed as shown. The deflection and stress of the structure at any instant during conditions shown in Figures 7.4a or 7.4b are pure shear deflection and stress only, and may be obtained using the elementary relations for shear deformation and stress. Although there are bending moments acting on the structure, during this period of transition, it is assumed that these do not develop bending stresses since the bending deformation has not as yet developed.

It should be noted that the shear loading described above and the resulting bending strain energy described in the next section during the period of overall effect are introduced solely as a means for determining the division of strain energy into symmetrical and skew-symmetrical parts. Having determined these, the total energy that the structure must absorb (as determined by the methods of Chapter 4) is divided into symmetric and skew-symmetric portions having the same ratio as the corresponding energies described above. Because the structure is assumed as a uniform free–free structure it will vibrate in symmetrical and skew-symmetrical modes. The division of the symmetrical energy among the symmetrical modes and of the skew-symmetrical energy among the skew-symmetrical modes will be approximated on the assumption that they are additive for each set and also (as will be seen shortly) on an assumed simplified triangular inertia loading based upon the deflection configurations of the symmetrical and skew-symmetrical modes.

PERIOD OF OVERALL EFFECT

Figure 7.4b shows the loading on the structure at the instant the 'period of overall effect' begins, due to an earthquake pulse load P applied at the base of the 'equivalent length' structure. *The loading shown is the equivalent (in the approximation used herein) to a suddenly applied load acting on the structure.* This is, perhaps, the key statement in the explanation of the proposed method, because the fact that the loading shown in Figure 7.4b is a suddenly applied loading, when used in conjunction with the physical requirement that the structure is a free–free beam (and that therefore the required balancing inertia loading remains constant and the same as shown in Figure 7.4b as long as the load P is acting on the structure), enables us to introduce the vibratory effects as being caused by a suddenly applied energy input of known value. This will now be described in greater detail.

The strain energy in a laterally loaded beam element, for gradually applied loads is (bending effects only considered)

$$U_{gal} = \int_0^l \frac{M^2 \, dx}{2EI} \tag{2}$$

By 'gradually applied' we mean that the loads are applied so slowly that the beam will be in deflection equilibrium as well as statical force equilibrium at all times. This is, of course, only a theoretically-attainable state but for all practical purposes a slowly applied load approximates the condition of gradually applied load.

However, the condition shown in Figure 7.4b is just that corresponding to a 'suddenly applied load' because the beam has not had time (it is assumed, and this very nearly so) to deform due to the bending effects. Under these circumstances the strain energy has the value

$$U_{sal} = \int_0^l \frac{M^2 \, dx}{EI} \tag{3}$$

Thus, at time $t = t_3$ with the loading shown in Figure 7.4b (which loading, as previously stated, remains constant), there has been an amount of strain energy introduced into the system of value

$$U = \int_0^l \frac{M^2 \, dx}{EI} \tag{4}$$

Of this strain energy,

$$\frac{U}{2} = \int_0^l \frac{M^2 \, dx}{2EI} \tag{5}$$

is required to account for the loading of Figure 7.4b which, therefore, corresponds to a quasi-static loading. The remaining energy

$$\frac{U}{2} = \int_0^l \frac{M^2 \, dx}{2EI} \tag{6}$$

goes into vibratory energy of the vibrating structure. It is just this energy which causes the structure to vibrate, and it vibrates in such a way that the sum of kinetic and strain energies is at all times equal to $U/2$ (less the amount damped from cycle to cycle).

In the present analysis, therefore, at all times $\geqslant t_3$, the condition on the structure due to a pulse loading is:

1. The quasi-static balancing inertia loading of Figure 7.4b exists.
2. The vibrating modes of Figure 7.2 exist and represent effects which superpose on those of (1). The amount of vibration (amplitude of oscillation) is determined by means of an energy balance: the total maximum strain energy of the vibrating structure is equal to $U/2$, which is just equal to the quasi-steady strain energy, or strain energy corresponding to the loading of Figure 7.4b.

Having described the state of the structural aystem during the period of overall effect, it is now necessary to explain how the vibrating strain energy is assumed apportioned among the different modes of vibration.

STRAIN ENERGY IN THE VIBRATING SYSTEM

The configurations assumed for the four modes are indicated in Figure 7.5 shown rotated through $90°$, for convenience. In this figure, the $k_n l$ terms refer to the free–free beam frequency solution

$$\omega_n = (k_n l)^2 (EI/ml^4)^{1/2} \text{ rad s}^{-1} \tag{7}$$

in which $m =$ mass per unit length.

The following should be noted in connection with these configurations:

1. The node positions are those obtained in the solution of the classical laterally loaded free–free vibrating uniform beam problem.
2. The maximum oscillation is assumed to be the same for each internal loop for any mode.
3. It is assumed that each loop may be approximated by triangles as shown in Figure 7.6 for S2 and AS3 only.

THE DEFORMATION OF THE VIBRATING STRUCTURE

The actual deformation mechanism for the structure is extremely complicated. It is initiated by the suddenly applied shear inertia loadings and consists of various (possibly) symmetric and anti-symmetric essentially non-superposable components deflecting from the vertical position. As an approximate approach, it will be assumed that the different modes are additive with each mode having its own period and frequency, and that the deformations start at time $t = 0$, from the vertical. If desired, the $t = 0$ time could be taken as t_3, the end of the period of transition (see Chapter 6).

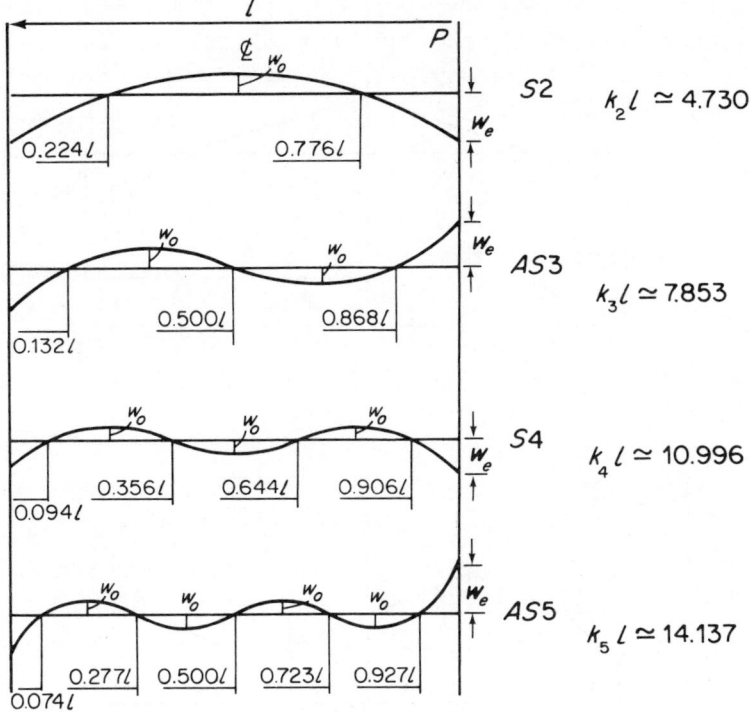

Figure 7.5

Consider typically, the S2 mode of vibration, in the maximum deflection position as shown in Figure 7.7. This will cause a deflection as shown in Figure 7.7(b), where zero deflection occurs at the points A, the nodes. w_{e_2} and w_{o_2} are assumed the loads on the structure causing the maximum deflection shown.

Because the structure (neglecting damping) vibrates on both sides of the vertical, the deflections δ_e and δ_o are *twice* the deflections that would be caused by gradually applied loads w_{e_2} and w_{o_2} applied to an initially undeformed vertical structure. Therefore the actual deflections from the vertical, due to the inertia loadings of Figure 7.7(a), are approximately one-half the 'gradually applied' values corresponding to this load.

The various zero-deflection positions (the nodes) are set as closely as possible to the frequency analysis positions for these as shown in Figure 7.5.

The following preliminary data is collected and assembled in preparation for the deformation and structural analysis. The code or specifications or governing agency or engineer set the earthquake magnitude, location, and other data from which, using the equations and curves of Chapter 4, the ground energy per unit area can be determined. Based upon known data or the designing engineers' expertise and judgement, values are assumed for the effective base area, Ae, and the effective length, l. In this connection, the author recommends the following until such time as sufficient data and experience are gathered to suggest otherwise:

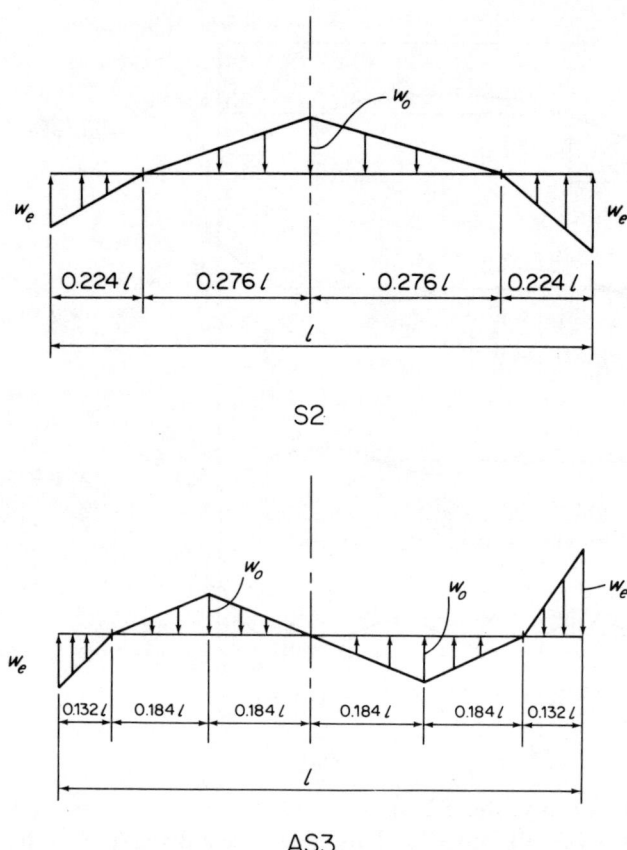

S2

AS3

Figure 7.6

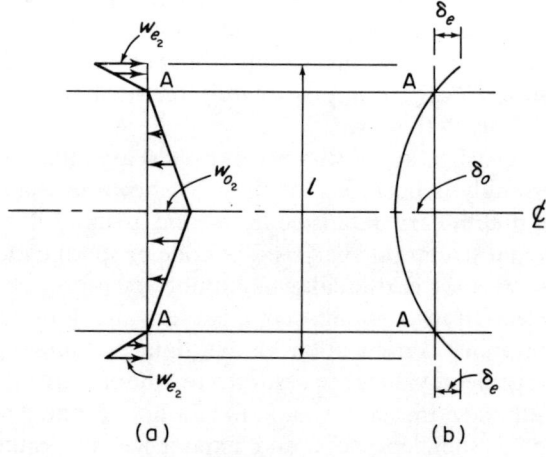

(a)   (b)

Figure 7.7

1. Assume a uni-material structure, with a single effective stiffness, EI.
2. For effective area, Ae, use the *overall* area of the structure base.
3. For effective length, l, use a value such that the two-node period of the structure (bending only considered) is equal to $0.1N$ as suggested by some approximate observations. N is the number of storeys in the structure and the period is determined from the free–free vibration solution, Figure 7.5.
4. For the low, squat (shear) structure use the actual length of the structure and consider all energy as shear energy except for specially-designed steel rigid frames with moment connections at the joints, and for large concentrated masses of equipment and similar items.

At this point, U_{total} is known, as are l, Ae, the period, and the dimensions, stiffness, etc.

THE DIVISION OF ENERGY — SEE TABLE 7.1

The division of energy within the vibrating structure is determined in the following sequence and approximate manner:

1. We shall assume bending strain energy (and dynamic energy) only are involved. Shear effects will be important for short, squat structures and during the period of transition for longer structures. See step 10 in the next section.
2. The total energy that the structure must absorb is $U_{total} = \varepsilon_{t_f}(Ae)$, and the timewise application of energy to the structure is as shown in Figure 4.3.
3. Because this energy is applied by the ground in an essentially smooth, continuous manner, with roughly equal alternating directional loads on the base, we shall assume that the quasi-static component of energy is not applied and all of the energy which the structure must absorb is taken in equal vibratory (dynamic) and potential (strain) energies. At any point in the vibrating cycle, the energy is equal to the sum of dynamic and strain energies. The structure is assumed to vibrate on both sides of a vertical.
4. The maximum strain energy occurs when the structure is at its maximum deflected position, which occurs on both sides of the vertical (neutral) position of the structure. At these locations the dynamic energy is zero, since the velocity of the structure is zero.
5. The maximum dynamic energy occurs when the moving structure passes through the neutral vertical position, corresponding to zero strain energy and maximum velocity.
6. The maximum strain energy (step 4) is equal to the maximum dynamic energy (step 5) of the same cycle, less the loss due to damping.
7. The total energy is damped by an amount ΔU per cycle.
8. Referring to Table 7.1, in which the various strain energy quantities are shown, we see that

$$\frac{U_{PS}}{U_{PAS}} = 6.82 \tag{8}$$

Table 7.1 Energy expressions

Sketch and loading	S = symmetric AS = Anti-symmetric EI = Constant	Energy, U $U_{TOTAL} = \varepsilon_{tf}(Ae)$	
Symmetric Component (P/2, P/2)		$U_{P_S} = 0.00406 \dfrac{P^2 l^3}{EI}$	
Anti-symmetric Component (P/2, P/2)		$U_{P_{AS}} = 0.00059 \dfrac{P^2 l^3}{EI}$	$\dfrac{U_{P_S}}{U_{P_{AS}}} = 6.82$ $\therefore U_S = 0.87\ U_{TOTAL}$ $U_{AS} = 0.13\ U_{TOTAL}$
S2 Component, $\dfrac{w_{e_2}}{w_{o_2}} = 1.23$		$U_S = 0.000259 \dfrac{w_{e_2}^2 l^5}{EI}$	CASE I U_{S2} and U_{AS_3} only $U_{S2} = 0.87\ U_{TOTAL}$ $U_{AS_3} = 0.13\ U_{TOTAL}$ $U_{S4} = 0$ $U_{AS_5} = 0$
AS3 Component, $\dfrac{w_{e_3}}{w_{o_3}} = 1.13$		$U_{AS_3} = 0.0000374 \dfrac{w e_3^2 l^5}{EI}$	CASE II $U_{S2}, U_{AS3}, U_{S4},$ and U_{AS5} $U_{S2} = 0.70\ U_{TOTAL}$ $U_{AS3} = 0.10\ U_{TOTAL}$ $U_{S4} = 0.17\ U_{TOTAL}$ $U_{AS5} = 0.03\ U_{TOTAL}$

Table 7.1 Energy expressions

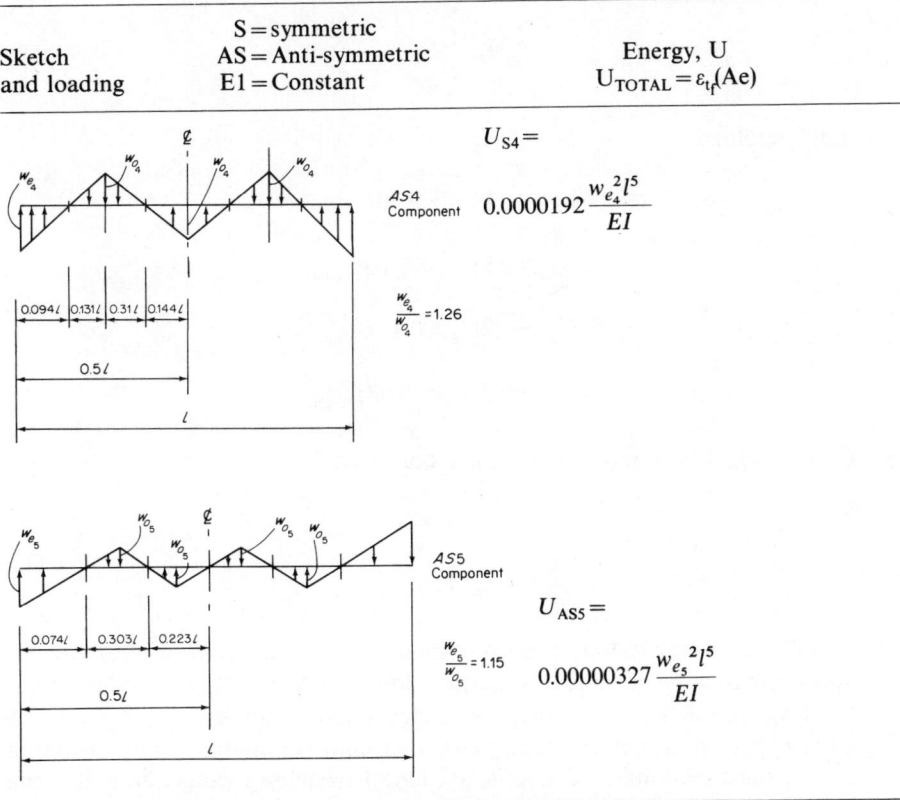

We assume this determines the ratio of total symmetric to total anti-symmetric strain–vibratory energies. That is,

$$\frac{U_{totalS}}{U_{totalAS}} = 6.82 \qquad (9)$$

From Eq. (2) above, we can now determine the total energies, U_{totalS} and $U_{totalAS}$ as follows

$$U_{totalS} = 0.87 U_{total}$$
$$U_{totalAS} = 0.13 U_{total} \qquad (10)$$

9. We consider only U_{S2}, U_{S4}, and U_{AS3}, U_{AS5} as the symmetric and anti-symmetric modes of vibration developed by the earthquake. Based upon the lengths of the loops developed and approximate strain and kinetic energy

analyses, we assume

$$\frac{U_{S2}}{U_{S4}} = \frac{U_{AS3}}{U_{AS5}} \cong 4 \tag{11}$$

and therefore,

$$U_{S2} \cong 0.8 U_S \cong 0.70 U_{total}$$
$$U_{AS3} \cong 0.8 U_{AS} \cong 0.10 U_{total}$$
$$U_{S4} \cong 0.2 U_S \cong 0.17 U_{total}$$
$$U_{AS5} \cong 0.2 U_{AS} \cong 0.03 U_{total} \tag{12}$$

Obviously, if only U_{S2} and U_{AS3} are considered,

$$U_{S2} \cong 0.87 U_{total}$$
$$U_{AS3} \cong 0.13 U_{total} \tag{13}$$

10. It will be assumed that the combined effect of all four modes can be approximated by a simple superposition of effects. This includes shear, bending, and deflection as these are determined by the loading of Table 7.1 and the periods as determined by either a frequency analysis or by assuming the constant section free–free, effective length frequency values. Note that the w_e and w_o values are determined from the known U_{total} quantity and the expressions in Table 7.1.

OUTLINE OF THE DESIGN-CHECKING PROCEDURE

We now outline, in step-wise form, a suggested design or checking procedure for a building using the methods and theory described. In this connection, it is emphasized that almost certainly refinements and modifications will be introduced as more experience is gained and additional data is collected. Also special structures such as bridges and dams may require different procedures, but in all cases these can be straightforward and logical extensions of the basic theory. In the next section a detailed application will be given for a particular structure.

As pointed out in the last section, the building has been located at a given point. The code or specifications or the engineers call for the design accelerogram or earthquake magnitudes and other data from which, assuming effective length and effective area are known, we can estimate the total energy which the structure must absorb. From Table 7.1 it is possible to determine the total energy for each mode of vibration.

Assume first the accelerogram is canonical. Then the energy can be assumed applied as either (a) a single total amount at $t=0$, (b) continuously in accordance with Eq. 7 of Chapter 4, or (c) a series of energy packets, whose sum equals the total energy, at some predetermined frequency.

It would appear that method (a) is the most conservative and will lead to the maximum stress and deflection condition. We shall assume this is so and will describe the structural analysis corresponding to this method of loading. Consider the two-node mode only. The others are handled similarly with the obvious time variations corresponding to their frequencies.

1. At $t=0$ (or $t=t_3$, see Chapter 6, the two-node free–free beam vibration is assumed to begin.
2. At the $\frac{1}{4}$ cycle, the maximum deflection is reached and the values of w_e and w_o are determined from the known energy value, $U_{S2_{total}} - U_{damped}$ in $\frac{1}{4}$ cycle, see Table 7.1.
3. At the $\frac{3}{4}$ cycle the maximum opposite deflection is reached and the values of w_e and w_o are determined from the *remaining* energy,

$$U_{S2_{total}} - U_{\text{damped in 3/4 cycle}}$$

again using the expressions in Table 7.1.
4. Continue thus, going from maximum strain energy position to maximum strain energy position, decreasing the energy from half cycle to half cycle due to damping and computing the corresponding w_e and w_o values, until all the energy has dissipated.
5. Do this for the U_{AS3}, U_{S4}, and U_{AS5} energies, Table 7.1, until all these energies have dissipated. Note that, because of the higher frequencies, these modes will dissipate to zero well before the U_{S2} mode, with U_{AS5} dissipating most quickly.
6. To find the approximate moment, shear, or deflection at any time, assume all four separate effects can be superposed, using times corresponding to the different frequencies.
7. If the energy is to be introduced in separate lump sums, as would be done, for example, for an accelerogram consisting of superposed canonical accelerograms, the same procedure is followed. We simply increase the amount of total energy at the appropriate times and calculate the corresponding w_e and w_o values at the maximum deflection positions of the vibrating structure in each of the four modes.
8. If only two modes, S2 and AS3, are used, the procedure is similar using the corresponding values of the energies, see Table 7.1.
9. If the structure contains significant (heavy) concentrated machine parts or similar objects, the energy which these absorb can be approximated from their locations in the structure and the corresponding mv^2 terms which can be estimated when the deformation is known. This energy must be subtracted from the total energy supplied by the ground. Or an equivalent distributed mass may be included in the mass of the structure.

10. If the structure is short and squat, then assume all the energy is absorbed as shear energy (less that amount taken by concentrated machines or by rigid frames in bending) and assume simple sliding motion back and forth from the initial undisturbed position.
11. In all of the foregoing, the maximum stress and deflection, for design purposes, should be less than specified code values. Failure will be indicated by excessively high values of stress and/or deflection.
12. If the building orientation with respect to a ray from the epicentre requires it, the ground energy may be decomposed into two perpendicular directions parallel to the principal axes of the structure and the response taken as a superposition of the two principal axes effects. This may be done because of the scalar identity $V^2 = V_x^2 + V_y^2$ and because the energy of the vibrating structure is given by the kinetic component $mV^2/2$. In the above, V = magnitude of velocity; V_x = magnitude of the velocity in the x principal direction; V_y = magnitude of the velocity in the y principal direction.

A TYPICAL DESIGN-ANALYSIS STRUCTURAL APPLICATION

Introduction

The previous section indicated, in general outline form, a suggested approximate design checking–analysis procedure for a building using the theory derived herein. In this section an actual application will be presented for a particular tall building. Most of the detailed calculations will be given. For those items which are best obtained using available (or easily prepared) computer programs, an outline only will be presented.

In a typical application, the code or specifications or the governing agency-engineer would specify:

1. Building dimensions, cross-section, materials.
2. Earthquake magnitude M, efficiency η, geology region $\mathscr{R}$.
3. Location of building, i.e. distance from epicentre.

From (2), we can determine,

4. $\Sigma\,(IS)_f$ and S_f and therefore the MID curve and ε_{t_f}, the total energy per unit area at the building location.

From (3) and (4) as a check we can determine the approximate intensity at the building location and therefore from Figure 4.5 approximate values of $\Sigma\,(a\,\Delta t)_f$ and t_f and also ε_{t_f}, the total energy per unit area which must be compatible with the value determined in (4) and (2). Thus in (1), (2), and (3) all of the earthquake information needed for the rational design analysis of the structure is available using the charts, equations and methods developed in the text.

In addition, we require test data values or the engineers' judgement determination of l, the effective length, and of Ae, the effective area of the structure. We shall indicate in some detail how the design analysis is performed for a typical assumed set of conditions (1), (2), and (3).

In this connection, from the known ε_{t_f} and (assumed) $\Sigma\,(a\,\Delta t)_f$ and t_f values, packets of energy could be applied to the structure at predetermined times by using Eq. 9 of Chapter 4, and the stress–deflection behaviour of the structure obtained for this assumed input of energy. However, it should be noted that according to Eq. 9 of Chapter 4 which is shown in Figure 4.3 of that chapter, a large part of the energy is introduced at small values of t. In other words, for practical design purposes, the assumption of total energy application at $t=0$ is probably not excessively conservative for a single canonical accelerogram.

Therefore, for our present purposes (and very likely for most actual design situations) it will be sufficient to make the conservative assumption that all of the energy is applied at time $t=0$, and furthermore this energy is apportioned to the S2 and AS3 modes only. The energy-packet design and the S4, AS5 contributions are handled in a similar manner if it is desirable or necessary to include these effects. For example, if a check on a particular actual response is desired, one could use the timewise variation of energy relation as well as the separate symmetric, anti-symmetric energy components and superposed canonical accelerograms, if desired. This will give a detailed representation of the stress and deflection response as determined by the present theory, including 'higher-order effects'.

It must be stated once more — the particular computations which follow are based upon the data available at this time. Various assumptions must be made (energy-magnitude, efficiency of earthquake, period of vibration, and others) and these will almost certainly be refined and will become more accurate as data are accumulated. The actual numerical results obtained must be evaluated and judged with this in mind. It is possible that these numbers are quite accurate. Or they may be somewhat approximate. The final determination of this must await the collection of additional experimental data, as is true in all fields of applied mechanics in which new approximate theories are introduced.

The technique which is used, however, is the one which will apply also when the additional data is at hand. Ultimately, it should be possible to prepare a 'handbook' in which all of the earthquake parameters are related to typical structures and the damage probability of structures is related to intensity and ε_{t_f}.

It will now be shown, in some detail, how the various design and theoretical charts and equations of Chapters 4, 6, and 7 are applied to a typical structural analysis. For the somewhat lengthy and complicated stress and deflection analyses, the general method only will be presented with a minimum of actual numerical computations. However, the detailed numerical solutions for these cases use elementary methods and are readily machine programmed.

The Design Analysis

The step-by-step procedure is as follows (the numbering of the actual computations will conform to these numbers):

1. The earthquake magnitude, geology, building dimensions, location, and materials of construction are specified.
2. From Figure 4.7, determine $\Sigma(IS)_f$ and S_f.
3. From Eq. 15 of Chapter 4, determine the total energy of the earthquake.
4. Assume an earthquake efficiency (specifications) and determine H, the total horizontal ground energy between S_i and $S_f = S_{III}$.
5. Using Eq. 14 of Chapter 4, determine U_{total}, the total horizontal ground energy to be absorbed by the building. Assume the effective area, Ae, in this step. Note, in effect, the efficiency of step 4 can be related to the assumed value of S_f/S_i of Eq. 14, Chapter 4, so that the choice of S_f/S_i for any particular case will be determined by accumulated data and experience and will not be arbitrary.
6. Determine E, I, and m for the structure. [Note: In this computation we shall assume a uni-material concrete structure. As noted earlier, experience may indicate that a more accurate analysis can be obtained using separate framing (i.e. steel) and outer shell (i.e. brick or concrete) structures.]
7. Compute the effective length, l, using the empirical relation

$$\text{period} = 0.1N \tag{15}$$

and the major mode equation for the free–free vibrating structure,

$$\omega = (4.73l)^2 \left[\frac{EI}{ml^4}\right]^{1/2} \text{ rad s}^{-1} \tag{16}$$

[Note: This introduces an approximation of the soil-foundation interaction effect. If, in the designing engineer's judgement, an adjustment is desirable, the value of l could be increased or decreased accordingly.]
8. As a rough accelerogram compatibility check, when sufficient information is available on ε_{t_f} as given on Figure 4.5 determine

$$\text{Total energy} = U_{total} = \varepsilon_{t_f} Ae \tag{17}$$

which should be approximately the same as the value obtained in step 5.
9. Determine U_{totalS} and $U_{totalAS}$ from

$$U_{totalS} = 0.87 U_{total}$$
$$U_{totalAS} 0.13 U_{total} \tag{18}$$

10. Assume

$$U_{S2} = U_{totalS}$$
and $\qquad\qquad\qquad\qquad\qquad\qquad\qquad\qquad\qquad\qquad\qquad$ (19)
$$U_{AS3} = U_{totalAS}$$

11. Determine w_{e_2}, w_{o_2}, w_{e_3}, and w_{o_3} using relations in Table 7.1.
12. Determine maximum shears and maximum moments using the loads of step 11.
13. Determine deflections using the loads of step 11.
14. Using an assumed damping factor, determine shear, bending, and deflection time histories.
15. If a combined canonical accelerogram is to be used, the total energies corresponding to each of the separate canonical curves will be applied to the structure at the times corresponding to the assumed accelerogram.
16. If the structure is a short, squat one (i.e. a shear structure) then all the energy is to be taken in simple sliding shear

$$U_{total} = k \frac{V^2 l}{2AG} \tag{20}$$

in which, initially it is recommended that l be taken as the actual overall length of the structure, k be taken as unity, and Ae be taken as equal to the overall area of the building. Determine V and check for the strength of the structure.

The analysis–computation corresponding to the above now follows:

1. We will consider an earthquake roughly equivalent to the Imperial Valley, 1940 quake. Our building will be the one previously used for illustrative purposes and it will be assumed that it is located at $S = S_f/2$. The specifications call for an earthquake magnitude about 6.5 in a region of assumed geology $\mathcal{R}_1$.
2. From Figure 4.7, this indicates that

$$S_f = 75 \text{ km}$$
$$\Sigma (IS)_f = 700 \tag{21}$$

3. Eq. 15 of Chapter 4 states

$$\text{Log}_{10} E = 11.4 + 1.5M$$

So that $\qquad E = 10^{21}$ ergs $\tag{22}$

4. If we assume an efficienty of 10 per cent, then

$$H = E\eta = 10^{20} \text{ ergs} \tag{23}$$

in which H is the total horizontal ground energy between S_i and S_f.

5. Using Eq. 14 of Chapter 4, noting the dimensions of the building is 100 ft × 100 ft ($=0.03$ km) and assuming this is the effective area, Ae, we can

determine the total energy to be absorbed by the building,

$$U_{total} = \frac{(0.03)}{2\pi S} \frac{H}{9} \left[\left(\frac{S_f}{S}\right)^{1/3} - \left(\frac{S_f}{S+0.03}\right)^{1/3} \right] \quad (24)$$

which, for $S_f = 75$ km, $S = 37.5$ km, $H = 10^{20}$ ergs, gives

$$U_{total} \cong 10^{12} \text{ ergs}$$
$$\cong 100{,}000 \text{ ft lbs} \quad (25)$$

as the total energy to be absorbed by the building.

6. The concrete outer and inner walls are assumed to be 12" thick. There are 12–24 WF 160 steel columns. We assume the neutral axis is at $d/3$, neglect tension in the concrete and assume the outside walls are 50 per cent effective because of window openings, etc. This is as shown in Figure 7.8. The principal axis of the building is assumed perpendicular to a ray from the epicentre.

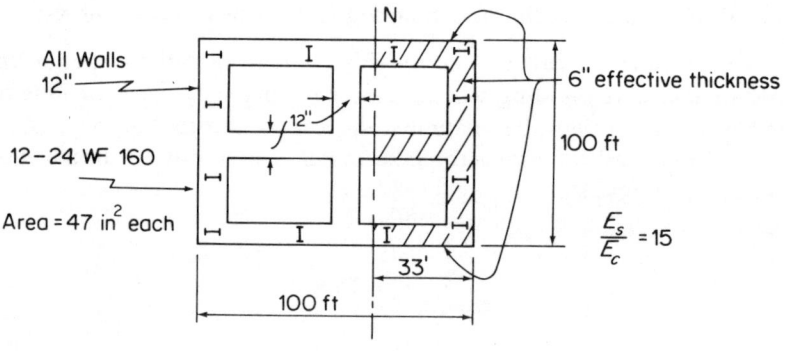

Figure 7.8

Using dimensions, weights, and similar data, we obtain weight of building, $W = 92{,}000$ lb ft^{-1}

$$I_{NA} = 186{,}000 \text{ ft}^4$$

$$E = 2{,}500{,}000 \text{ (144) psf} \quad (26)$$

$$A_{eq} = 640 \text{ sq ft} = \text{equivalent concrete area}$$

$$m = \frac{W}{g} = \frac{92{,}000}{32.2}$$

7. The empirical relation for the period of the building

$$\text{period} = 0.01N \text{ s} \qquad (27)$$

gives, for $N = 40$ storeys,

$$\text{period} = 4 \text{ s for a complete cycle} \qquad (28)$$

or

$$\omega \cong 1.5 \text{ radians s}^{-1}$$

Therefore, from Eq. 16, we have

$$l^3 = (4.73)^2 \left[\frac{EI}{W/g} \right]^{1/2} \qquad (29)$$

and using the values in step 6,

$$l = 1200 \text{ ft, say} \qquad (30)$$

which indicates that the soil–interaction effect is assumed as approximately accounted for as shown in Figure 7.9.

As pointed out, if this value of l appears to warrant modification, this could be done by the designer.
8. This check awaits additional data on the tentative Figure 4.5.
9. Using the relations as given in Table 7.1 and the value of U_{total} from step 5, we determine

$$U_{totalS} = 0.87 U_{total} = 87{,}000 \text{ ft lb}$$
$$U_{totalAS} = 0.13 U_{total} = 13{,}000 \text{ ft lb} \qquad (31)$$

10.

$$U_{S2} = 87{,}000 \text{ ft lb}$$
$$U_{AS3} = 13{,}000 \text{ ft lb} \qquad (32)$$

11. From Table 7.1,

$$87{,}000 = \frac{0.000259 w_{e_2}^2 l^5}{EI}$$
$$13{,}000 = \frac{0.0000374 w_{e_3}^2 l^5}{EI} \qquad (33)$$

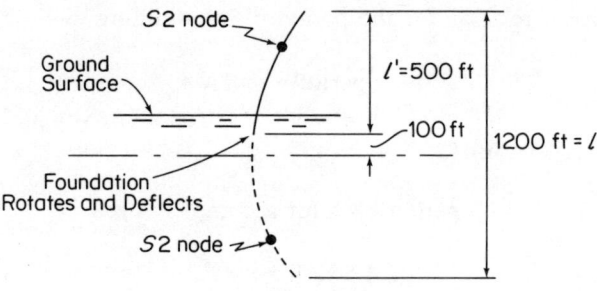

Figure 7.9

$$\frac{w_{e_2}}{w_{o_2}} = 1.23 \tag{34}$$

$$\frac{w_{e_3}}{w_{o_3}} = 1.13$$

so that,

$$w_{e_2} = 3100 \text{ lb ft}^{-1}$$
$$w_{o_2} = 2500 \text{ lb ft}^{-1}$$
$$w_{e_3} = 3000 \text{ lb ft}^{-1} \tag{35}$$
$$w_{o_3} = 2650 \text{ lb ft}^{-1}$$

12. The ratio of

$$\frac{\omega_{AS}}{\omega_S} = \frac{61.6}{22.4} \sim 3 \tag{36}$$

Therefore the w_2 and w_3 will combine to give the approximate maximum superposed effect when the symmetrical vibration is at the $\frac{1}{4}$ cycle and the anti-symmetrical vibration is at the $\frac{3}{4}$ cycle. The loadings will then be approximately as shown in Figure 7.10, and approximate values of maximum shear and maximum moments (due to the vibratory inertia loads) are

$$V_{1-1} = w_{e_2}(0.112l) + w_{e_3}(0.066l)$$
$$= [3100(0.112) + 3000(0.066)](1200) \tag{38}$$
$$= 650{,}000 \text{ lb}$$

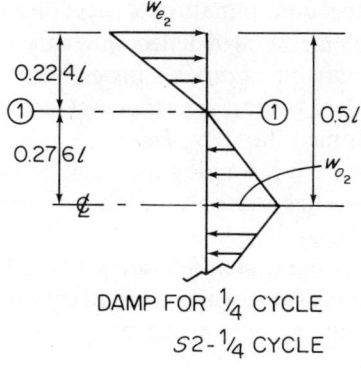
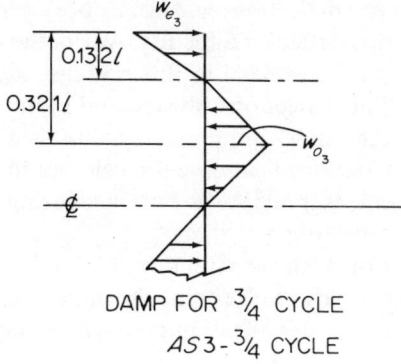

Figure 7.10

and

$$M_{1-1} = 545(1200)^2(0.15)$$
$$= 117{,}700{,}000 \text{ ft lb} \qquad (39)$$

or

$M_{\text{\textcentoldstyle}} =$ Approx. overturning moment

$$= 347(0.43)(l)^2 - 2500(0.276)l(0.092)l \qquad (40)$$
$$= 3.17(0.43)(1200)^2 - 2500(0.276)(0.092)(1200)^2$$
$$= 125{,}000{,}000 \text{ ft lb}$$

Then

$$\tau_{max} = \text{shear stress} = \frac{V}{A_{EQ}} \qquad (41)$$
$$\cong \frac{650{,}000}{640} \cong 1000 \text{ psf}$$

$$\sigma_{max 1-1} = \frac{MC}{I} = \frac{(117{,}700{,}000)(30)}{186{,}000}$$
$$\cong 200{,}000 \text{ psf} \qquad (42)$$
$$\cong 1400 \text{ psi}$$

The above values will be modified by the effective eccentricity of the dead weight of the building, which will be dependent upon the actual spandrel and connection construction details.

13. The deflections will not be obtained. The actual determination of these offers no particular difficulty. Any of the elementary methods of deflection analysis can be utilized in this exercise, including available computer programs.
14. The deflections, shears, and moments may then be obtained timewise using the known frequency equations and an assumed damping factor.
15. Combined canonical accelerograms would require a simple superposition of effects previously obtained using the known (assumed) time phases and accelerogram shapes.
16. For a shear structure, say a building 15 ft high, steel columns 8WF31, everything else being the same as the structure of step 6, then the total energy, $U_{total} = 100,000$ ft lb is taken in shear strain energy and we have

$$U_{total} = \frac{kV^2 l}{2AG} \tag{43}$$

with $k = 1$, assumed
$l = 15$ ft
$A = 12(9.1)(15) + 400 = 2000$ sq ft, equivalent concrete
$G = 1,000,000(144)$ psf,

So that

$$\tau = \frac{V}{A} = \text{Shear stress}$$

$$= \left[\frac{2(1,000,000)(144)(100,000)}{(2000)(15)}\right]^{1/2}$$

$$= 220 \text{ psi}$$

which indicates that concrete stresses are high enough to be a cause for concern, particularly at corners where stress concentrations occur and cracks frequently initiate.

The above represents a typical macroscopic or overall analysis for a particular structure.

Following are some observations concerning connections, details of connections and similar items. According to a fundamental premise of this book, namely that energy is the earthquake effect that must be designed for, the most desirable characteristic of all beam connections and similar details is that they be capable of absorbing large amounts of energy, consistent with their ability to resist the imposed stresses. Friction, for example, as evidenced by tight, contigious, moving surfaces is an effect that should be exploited. In all, the sum of the energy absorbing details enters into the 'damping' factor used for the vibrating structure. The optimum design of energy absorbing connection details represents an important aspect of the overall structural analysis in earthquake engineering.

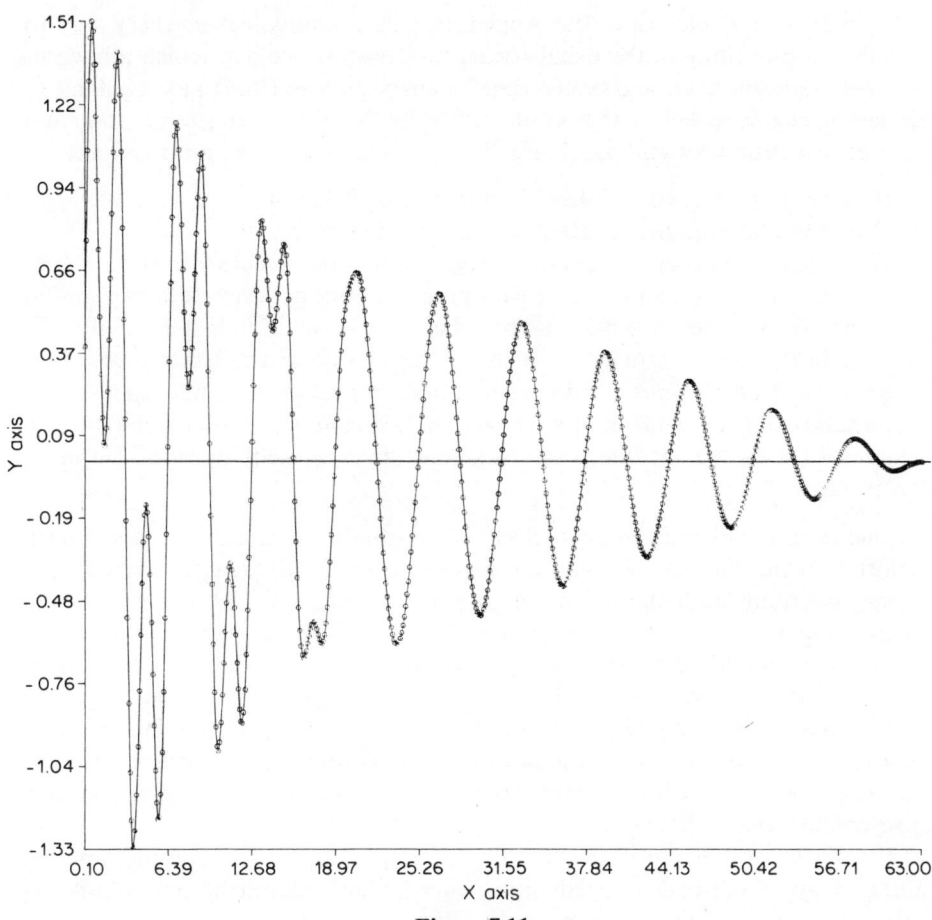

Figure 7.11

Figure 7.11 represents a typical superposition for deflection of a typical horizontal plane in the structure, considering the S2 and AS3 components only with damping equal to 0.1. The moment or shear at any point would be determined in the same way, using the w_e and w_o loadings, also damped.

CONCLUSION

It is essential that base values (calibrations) be determined for the various energy relations. New, fundamental procedures have been recommended in this chapter but, as pointed out, these are subject to modification and revision as data are accumulated concerning the energy–intensity correlations. The basic theory, the fundamental ideas which integrate the accelerogram (temporal energy)–isoseismal (spacewise energy) concepts still apply. What is needed are the experimental data that will permit one to use the methods with a greater degree of confidence and certainty.

As shown in Table A2 of the Appendix, this is equivalent in every way to conditions prevailing in the usual structural design process in which allowable stresses, allowable loads and similar data had to be gathered and made available to designing engineers before they could apply the theories of stress and strain and strength to structures and machines. To be more specific, we need to know

1. How to determine the effective base areas for different cases.
2. How to determine the effective length for different cases.
3. How the total energy applied to — say — a building structure is absorbed by the separate parts such as the steel framing, the brick or concrete facing, and so on. Do we assume a uni-material based upon relative moduli of elasticity as is done in elementary strength of materials and reinforced concrete theory? Or do we separate the frame and steel (and interior machines, furniture and other objects) using a partial energy for each including an mv^2 term for the interior items? Or do we include these last items as equivalent parts of the main structure?

One technique which suggests itself is to formulate a handbook of standard solutions using various combinations and values of effective lengths and effective areas, assuming both the uni-material and separate material approaches. By correlating the solutions so obtained with actual damage or with intensity numbers, it should be possible to come up with a set of 'code' or 'specification' recommendations which designers could use as starting points in their earthquake engineering analyses. As noted earlier, every solution obtained, every analysis of earthquakes, will increase our store and fund of data and will permit us to use the rational methods derived in the text with greater and greater accuracy and confidence.

Finally, although the discussion and example used refer to buildings, the methods and theories developed apply (when suitably modified) to non-building structures such as bridges, dams and others.

Appendix

Earthquake Engineering and Applied Mechanics

In this Appendix some further observations are made dealing with the basic content and form of the book. These will include statements concerning the place of earthquake engineering as a subsection of 'applied mechanics'. The following paragraph is a relevant introduction to the discussion.

In late 1981, a group of leading geologists, seismologists, and engineers gathered in Knoxville to discuss earthquakes within the eastern United States. (*Earthquakes and Earthquake Engineering; The Eastern United States*, Sept. 1981, Knoxville, Tennessee, Proceedings available from Earthquake Engineering Research Institute, 2620 Telegraph Ave., Berkeley, California 94704.) Richard A. Kerr, in *Science*, 9 October, 1981, summarized papers and talks presented. Quoting from his report, 'Henry Degenkolb of H.J. Degenkolb Associates..., a prominent seismic engineer for many years, held a sobering slide show during the meeting's final panel discussion... He showed the (1973) code-designed Imperial County Services Building that was torn down as a total loss following a moderate southern California earthquake in 1979 because its stylish open ground floor weakened it (*Science*, 29 August, 1980, p. 1006). The old masonry courthouse across the street suffered no damage. He showed several buildings in downtown Managua after the great earthquake there. All were still standing, but the one designed to resist the greatest shaking suffered the most damage, while the one with no seismic design suffered the least damage. 'Doesn't this tell us that perhaps we're on the wrong track?' Degenkolb said. 'The real guts of earthquake engineering is not contained in present codes. We don't fully understand the tie-in between what we measure (severity of shaking) and damage'.

This book presents a new and different 'track' dealing with the 'guts' of earthquake structural engineering. 'Shaking' is not the critical design parameter; 'energy' is the determining quantity.

Practically all current approaches to earthquake engineering analysis are straightforward applications and extensions of classical vibration and elasticity theory. The point of view taken herein is that the earthquake engineering phenomenon is a completely separate discipline in applied mechanics, just as elasticity and heat flow and others are separate fields. Therefore earthquake engineering has its own parameters, variables, constants, equations, invariants, and similar unique expressions.

A major corollary aim therefore was to uncover these quantities in the forms appropriate for engineering structural design purposes. It was expected that they might be different from the more familiar engineering terms and this is, in fact, the case.

In effect, it was assumed that insofar as earthquake engineering structural design was concerned, the fundamental engineering design method must find its source in the two sets of earthquake field (experimental) observables. These two basic sources of primary data which are the fountainheads of the methods described are

1. The accelerogram record.
2. The isoseismal (intensity) contour chart.

Invariants were looked for, these being quantities or relations that, *subject to the reality of the experimental accuracy*, appear to be approximately constant for all earthquakes, wherever and whenever they occur. Furthermore, these should be in a form that ultimately lends itself to structural engineering design applications — in as *simple* and *useful* a manner as possible — because engineers are practical workers in the market place and experience, over many years, indicates that the more complicated the analysis, the more suspect it is and the more likely it is to have errors and inconsistencies. Which is not to say that 'simplicity' itself is the be-all and end-all of analysis. However a reliance on physical reasoning (properly modified with engineering experience and intuition) does have a place in the more complicated engineering analyses, of which earthquake engineering surely is an outstanding example.

The situation being what it is in earthquake engineering, it was obvious that approximations of various orders were inevitable. But at some points the degrees of approximation would have to be tested, as indicated in the text.

The theory as developed and presented in the book is based upon field data, dimensional homogeneity, physical reasoning, and engineering experience and intuition. It assumes a 'canonical' form for the accelerogram (which many do, in fact, have approximately and those that do not can generally be represented by a superposition of canonical forms) and that the contours of the isoseismal map are circular which, again, is approximately true for many isoseismal maps.

Following this the various parameters and invariants are utilized including (in addition to the basic accelerogram and isoseismal parameters) the magnitude of the earthquake M, the efficiency, η, and various temporal and spacewise energy variation terms. Finally, and not the least important, there is the 'geology' which is related to focal depth as well as to the geology in a broad context and this is accounted for, at this time, by three different 'regions', $\mathscr{R}_{1,\ 2}$, and $\mathscr{R}_3$. If more regions are needed, they can be introduced.

All of the above are as shown on the different charts summarized in Chapter 4. There may be other charts that follow from the theory that will be useful in the design process but these will have to await the determination of actual numerical

Table A.1 Comparison between the theoretical formulations of linear elasticity and earthquake engineering

	Linear elasticity	Earthquake engineering
Basic measured quantities	Stress tensor	$\Sigma\,(a\,\Delta t)$, t (acceleration index, time)
	Strain tensor	$\Sigma\,(IS)$, S (intensity index, distance)
Restraints	Boundary conditions	Initial and final conditions
Measured experimental physical constants	E modulus of elasticity v Poisson ratio	$\mathscr{R}$ the geology S_f, $\Sigma\,(IS)_f$ t_f, $\Sigma\,(a\,\Delta t)_f$ final values
Compatibility conditions	In terms of strain or stress	Various energy charts Intensity vs. $\Sigma\,(a\,\Delta t)_f$, t_f, geology M vs. $\Sigma\,(IS)_f$, S_f, geology
Equations connecting the basic quantities	Hooke's Law (invariant)	$\Sigma\,(a\,\Delta t)$ vs. t relation $\Sigma\,(IS)$ vs. S relation (invariants)
Other relations or equations	Equilibrium	Time and spacewise variations of ground energy M-energy equation

quantities for various earthquakes under different conditions.

In accordance with established engineering practice, code values may be assigned for the various terms considered, including factors of safety to account for the inherent uncertainty of the available data, and the entire procedure can be programmed and used in consulting design offices.

The tabulations which follow compare earthquake engineering as developed herein with two of the familiar disciplines that are part of applied mechanics. The first, Table A.1, deals with the theoretical formulations of earthquake engineering and linearized elasticity. In the table, the correspondence is shown between the two sets of basic terms, equations, invariants, and similar quantities indicating the basic similarities and differences in the two formulations. In the second, Table A.2, a comparison is made between the *application* methods of the book's earthquake engineering theory and the *application* methods of a typical linearized elasticity (commonly called structural engineering) problem.

Table A.2 shows that a number of the design parameters must be assumed. Initially, the source of this data will be limited so that a greater burden will be placed upon the experience and judgement of the designing engineer. As more and more studies and designs are made, comparisons between design and actual behaviour will be possible and the code–field databases will become more certain and more accurate for engineering use.

Table A.2 Comparison between application methods of earthquake engineering and ordinary structural engineering

Design Data bases	Ordinary structural engineering design-analysis	Earthquake engineering design-analysis
Assumed known code or specification data	Floor loadings, dimensions and materials wind load, impact load and similar data	t_f = maximum time for canonical accelerogram based on existing accelerograms M = magnitude of earthquake η = efficiency of earthquake $\mathscr{R}$ = geology S = location (and orientation of structure) δ = damping factor Dimensions and materials of structure
Known from experimental or field data or must be assumed	Allowable pile loads, soil conditions, allowable stresses for materials and similar data	M–energy relation (empirical — available) ε_{t_f}–intensity relation M–$\Sigma\,(IS)_f$–S_f relation (experimental — partly available) W/H–S/S_f relation (theoretical — available, to be verified) $\varepsilon_t/\varepsilon_{t_f}$–$t/t_f$ relation (theoretical — available, to be verified) Ae = equivalent area l = equivalent length

To summarize, and in conclusion, it must be emphasized that (a) even if the accelerograms are *not* canonical or not representable by superposition, and (b) even if the isoseismal contours are *not* circular, these assumptions can reasonably be made with as much certainty as otherwise in view of the manner in which the data are determined. In other words, the assumptions are *not* unreasonable from the engineering point of view. Also, the invariants which follow from the assumptions have been verified for a number of earthquakes that have occurred all over the world during the past 500 years. The energy relations (temporal and spacewise) also are physically and technologically reasonable and the entire development does then lead to a straightforward rational approximate design method which — when the design parameters have been firmly established — will enable the average engineering design office to design a structure for an earthquake loading with an acceptable degree of engineering accuracy and confidence using a procedure which includes all of the major earthquake engineering parameters that one should expect to be included in such a design. A similar statement applies for damage assessment analyses.

There will surely be some anomalous behaviours. There are 'renegade' phenomena in every field — unusual, unexpected, or unexplainable occurrences that are beyond the scope of known rational analyses. It can only be hoped that these will not occur or if they do that the design used will have a sufficient factor of safety to take care of any possible excessive stressing or tendency to fail. If they occur (and they almost certainly will occur because of localized pockets of special soil or geologic conditions), the data they generate will be incorporated into the new theory and will thereby advance our understanding of the fundamental earthquake engineering processes.

INDEX

I – EARTHQUAKES REFERRED TO

Agadir (1960), 40, 52, 53
Arboledos (1950), 40, 52

Bantarkawing (1971), 40, 52
Benavente (1909), 40, 52
Bucharest (1977), 17, 18, 27, 51

Charleston (1886), 2, 40, 52
Corinth (1981), 40, 52, 53

Faial (1926), 40, 52
Friuli (1976), 22, 33, 38, 40, 52

Hawke's Bay (1921), 40, 52

Imperial Valley (1940), 33, 39, 40, 52, 95

Lice (1975), 40, 52, 53
Lima (1966), 17, 18, 25, 51
Lisbon (1755), 40, 52

Madeira (1941), 40, 52
Madjene (1969), 40, 52
Managua (1973), 103
Messina (1908), 40, 52
Mexico (1962), 40, 52

Orleanville (1954), 40, 52, 53

Romania (1977), 22

S. Jorge (1757), 40, 52, 53
S. Miguel (1522), 40, 52
San Fernando (1971), 17, 18, 26, 33, 38, 40, 51, 52, 53
San Francisco (1906), 40, 52, 53

Taft (1952), 17, 18, 24, 51
Tangshan (1976), 40, 52, 53
Tolmezzo (1976), 17, 18, 23, 51

Udine (1873), 31, 33, 39, 40, 52

Valparaiso (1906), 40, 52

Wairoa (1932), 40, 52
Washington State (1872), 33, 40, 52
Wewak (1946), 40, 52

II – AUTHOR INDEX

Armbruster, J., 2
Austurias, J., 36

Bath, M., 37
Benioff, H., 13
Berlin, G. L., 37
Bolt, B. A., 13, 23, 35
Bullen, K. E., 37

Degenkolb, Henry, 103
Donovan, N. C., 37

Espinosa, A. F., 31, 36

Fattal, 22

Gayhart, E. L., 76
Gere, J. M., 37
Giorgetti, F., 37

Idress, I. M., 22

Kerr, R. A., 13, 103

Machado, F., 37
McLachlan, N. W., 76
Mott, N. F., 8, 13

Neumann, F., 37, 38
Newmark, N., 1, 13, 28, 36, 56, 76

Quesada, A., 36

Ramirez, E. J., 37
Richter, C. F., 28
Rosenblueth, E., 1, 13, 28, 36, 56, 76

Seeber, L., 2
Shah, H. C., 37
Steinbrugge, K. V., 37

Wiegel, R. L., 22, 37, 76

Zencr, C., 76

III – TOPICS

accelerogram, xii, 30
  canonical, 14, 42
  index, xii, 42
  invariant, Chap. 2, 42
accuracy in earthquake engineering, 41
Alpide Belt, 48
American Iron and Steel Institute, 37
applied mechanics and earthquake engineering, 103
Australian Congress Earthquake Engineering Symposium, 37

base area, effective or equivalent, 59, 63, 64, 76, 87
bending energy, 87

canonical accelerogram, 14, 42
circular approximation for isoseisms, 32
Circumpacific Belt, 48
CNEN-ENEL, 22
Cornell University high pressure research, 11

damage assessment, xii, 28–30, 41, Chap. 5
damping, 63, 67, 87, 95, 101
dead material, 6, 8, 9, 10
deep focus earthquake, xii, Chap. 1, 14
deflections of structure, 85, 101
Department of Interior (USGS), xi
design-analysis, typical, 92, 94
design charts, Chap. 4
design-checking procedure, 90
division of energy, 87
dynamic energy, 87

efficiency of earthquake, xii, 41, 54 f, 94
effective or equivalent base area, 59, 85, 87, 94

effective or equivalent length, 59, 67, 87, 94
elasticity and earthquake engineering, 103, 105
energy compatibility, 54 f, 59, 105
energy loss, damping, 67
energy of earthquake, xii, 9, 11, 51, 54, 55, 58, 78
  strain, 9, 10
energy per unit area, 59
energy variation
  spacewise, 45, 55, 56, 58, 78
  temporal, 44, 56, 78

fault slippage, 1
fracture, 3
  stress, 5, 12
free-free beam, 78
frequency effect, xii, 14, 50

geological (soil) condition, 35
geology, xii, 14, 20, 48, 54

healing of earthquake, 11

Imperial County Services Building, 103
intensity, 28, 30, 49, Chap. 5
  index, 32
intensity accelerogram, geology damage chart, 50, 58
invariant
  accelerogram, Chap. 2, 42, 57
  isoseismal, Chap. 3, 43, 58
isoseismal, xi, 30, 43, 55
  index, xii, 21 f, 51
  invariant, Chap. 3

kinetic energy, 67

Lamont–Doherty Geological Observatory, 2
length, effective or equivalent, 59, 63, 64, 78, 87
  of time effect, 63, 65
Low Seismicity Region, 48

magnitude–energy relation, 58
magnitude–intensity–distance chart, 53, 54, 58
magnitude isoseismal correlation, 51, 58
magnitude of earthquake, xii, 41, 51, 53, 54, 78
mechanism, earthquake, xii, Chap. 1, 30
MID chart, 53, 54, 58

model–prototype, xii
model–prototype relations, 63, 68
model scaling requirements, 68
modes of vibration, 85
Modified Mercalli (MM) Scale, 28, 57

National Academy of Sciences, 37
National Science Foundation, xi
nodes of vibrating structure, 85

outer reach of effect, 4, 6, 8

Pacoima Dam, 19
period of localized effect, 66
period of overall effect, 66, 80, 82
period of transition, 66, 80, 81
period of vibration of structure, 94
phase change in earthquake mechanism, 1, 3, 10
Plains Region $\mathscr{R}_3$, 48
potential (strain) energy, 67
prediction of earthquake, 11
P waves, 10

ratio of effects in scaling analyses, 69
response of structure, 79
Richter scale, 12
rupture of earthquake mechanism, 3

scaling requirements, 68
shear structure, 87, 95
similarity solution, 7
soil-foundation interaction, xii, 67
soil (geological) condition, 35
spacewise variation of surface horizontal energy over the field, 45

special topics in earthquake engineering, Chap. 6
standard structures, 59
strain energy of mechanism, 10
strain energy of structure, 67, 83, 84, 87
stress compatibility of mechanism, 9
stress in structure, 100
structural analysis procedure, Chap. 7
structural design in earthquake engineering, xii
structural engineering and earthquake engineering, 105, 106
suddenly applied load, 82
superplasticity of the earthquake mechanism, 4
Surface Fault Region $\mathscr{R}_1$, 48
Surface Horizontal Energy (SHE), 15, 20, 21, 30, 35, 44, 45
S waves, 10
symmetry–anti-symmetry–unsymmetry, Chap. 7

tectonic earthquake, 1
tearing stress of earthquake mechanism, 5, 12
timewise variation of surface horizontal energy at a point, 21, 78
trigger of earthquake mechanism, 11, 12

uniqueness-existence hypothesis
   accelerogram invariant, 17, 20
   isoseismal invariant, 34

velocity of a small disturbance, 5, 65